Tabea Gisela Schwark

Deformation and Fracture Properties of the Soft Magnetic Composite Somaloy 700 3P on Different Length Scales

Schriftenreihe
des Instituts für Angewandte Materialien
Band 74

Karlsruher Institut für Technologie (KIT)
Institut für Angewandte Materialien (IAM)

Eine Übersicht aller bisher in dieser Schriftenreihe erschienenen Bände finden Sie am Ende des Buches.

Deformation and Fracture Properties of the Soft Magnetic Composite Somaloy 700 3P on Different Length Scales

by
Tabea Gisela Schwark

KIT Scientific Publishing

Dissertation, Karlsruher Institut für Technologie
KIT-Fakultät für Maschinenbau

Tag der mündlichen Prüfung: 21. September 2017
Erster Gutachter: Prof. Dr. rer. nat. Oliver Kraft
Zweiter Gutachter: Prof. Dr. rer. nat. Michael J. Hoffmann

Impressum

KIT Scientific Publishing

Karlsruher Institut für Technologie (KIT)
KIT Scientific Publishing
Straße am Forum 2
D-76131 Karlsruhe

www.ksp.kit.edu

Print on Demand 2018 – Gedruckt auf FSC-zertifiziertem Papier

ISSN 2192-9963
ISBN 978-3-7315-0759-8
DOI 10.5445/KSP/1000080372

Deformation and Fracture Properties of the Soft Magnetic Composite Somaloy 700 3P on Different Length Scales

zur Erlangung des akademischen Grades eines

Doktors der Ingenieurwissenschaften

von der Fakultät für Maschinenbau
des Karlsruher Instituts für Technologie (KIT)

genehmigte

Dissertation

von

Tabea Gisela Schwark

aus Saarlouis

Tag der mündlichen Prüfung: 21.09.2017
Erster Gutachter: Prof. Dr. rer. nat. Oliver Kraft
Zweiter Gutachter: Prof. Dr. rer. nat. Michael J. Hoffmann

Kurzfassung

Die Verwendung von weichmagnetischen Verbundwerkstoffen (SMC) in Geräten zur Energieumwandlung bietet eine hohe magnetische Permeabilität, einen hohen Widerstand und damit geringe Wirbelstromverluste sowie ein geringes Bauteilgewicht und eine geringere Bauteilgröße. Diese einzigartigen Eigenschaften der SMCs führen zu neuen Designoptionen für 3D-Komponenten, die neben den magnetischen Eigenschaften der SMCs jedoch auch eine gewisse mechanische Robustheit des Materials verlangen.

Typischerweise bestehen SMCs aus großen Eisenpartikeln, die mit einer relativ dünnen anorganischen Schicht ummantelt sind. Diese Isolierschicht führt zu einem höheren spezifischen Widerstand, wodurch die Wirbelstromverluste reduziert werden. Die Eisenpartikel dominieren in vorteilhafter Weise die magnetischen Eigenschaften. Die Kombination von weichen Eisenpartikeln mit einer spröden Isolierschicht bewirkt jedoch ein eher schlechtes mechanisches Verhalten der SMCs.

In der vorliegenden Arbeit wurde Somaloy 700 3P untersucht, das aus reinen Eisenpartikeln besteht, die mit einer dünnen anorganischen Phosphorousschicht ummantelt sind und die nach der Verdichtung wärmebehandelt werden, um eine festigkeitssteigernde Oxidschicht zu bilden. Transmissionselektronenmikroskopische Bilder zeigen, dass die Partikelgrenzen je nach Komplexität ihrer Schichten in vier verschiedene Typen klassifiziert werden können. Unter Verwendung eines fokussierten Ionenstrahles wurden Mikrodrucksäulen, Mikrobiegebalken und Mikrozugproben gefertigt, mit je nur eine Grenze oder nur Eisenpartikeln, und in einen Nanoindenter und beim Zugversuch mit einer Mikrozugapparatur getestet wurden. Zusätzlich wurde die Nanoindentation sowohl an Partikeln als auch an Partikelgrenzen durchgeführt. Auf der Makroskala wurden Vierpunkt-Biegeversuche durchgeführt, um die Bruchzähigkeit der SMCs zu bestimmen. Alle Tests auf Mikro- und Makroskala zeigten, dass die Grenzfläche zwischen Partikel und Partikelgrenze kritisch für das Versagen ist. Die Grenzschichten scheinen hart, aber duktil zu sein. Beispielsweise zeighten die sogenannten Typ-4-Partikelgrenzen ein einzigartiges mechanisches Verhalten aufgrund ihrer zusätzlichen porösen Eisenoxidschicht, wodurch die Grenzfläche im Vergleich zu den spröden Partikelgrenzflächen der anderen Grenztypen duktil verformt wurde. Es wurde auch gezeigt, dass mikromechanische Tests erfolgreich an komplexen Materialien wie den SMCs durchgeführt werden können.

Abstract

The use of Soft Magnetic Composites (SMC) in devices for energy conversion offers high magnetic permeability, high resistivity, and thus small eddy current losses as well as reduced weight and size of devices. These unique properties of the SMCs lead to new design options for 3D components, which demand a certain mechanical robustness of the material besides the magnetic properties of the SMCs.

Typically, SMCs consist of large iron particles coated with a fairly thin inorganic layer. This insulating layer leads to an increase in resistivity reducing eddy current losses. The iron particles dominate in a favourable manner the magnetic properties. The combination of soft iron particles with a brittle insulating layer causes, however, a rather poor mechanical behaviour of the SMCs.

In this work, the commercially available Somaloy 700 3P was studied, which consists of pure iron particles that are coated by a thin inorganic, phosphorous layer, and which is annealed after compaction in order to form a strength-enhancing oxide layer. Transmission Electron Microscopy images show that the boundaries can be classified into four different types according to the complexity of their layers. Using focused ion beam preparation, micropillars, microcantilevers, and microtensile test specimens, containing only one boundary or only bulk iron particles, were produced and tested by a nanoindenter and a microtensile test setup. Additionally, nanoindentation was performed on both particles and boundaries. On the macroscale, four-point bending tests were conducted to determine the fracture toughness of the SMCs. All tests on both micro- and macroscale showed that the particle-boundary interface is critical in terms of failure. The boundary layers seem to be hard but ductile. The type-4 boundary showed unique mechanical behaviour due to its additional porous iron oxide layer making the particle-boundary interface ductile compared to the brittle particle-boundary interfaces of the other types of boundaries. It was also shown that micromechanical tests can be successfully applied on complex materials such as the SMCs.

Content

List of Abbreviations

a	Characteristic size of a defect / crack length
A	Area of the pillar at the top
a_c	Critical defect size
AC	Alternating current
a_{max}	Largest value of crack lengths a_1, a_2 and a_3
a_{min}	Smallest value of crack lengths a_1, a_2 and a_3
ASTM	American Society for Testing and Materials
b	Width of the cantilever
B	Magnetic flux density / magnetic induction
B_b	Width of the four-point bending specimen
B_s	Magnetic saturation (flux density)
C	Compliance
CCS	Cleaning Cross Section
CSM	Continuous stiffness measurement
d	Displacement of cantilever at loading point
DC	Direct current
DCM	Direct contact module
DIN	German Institute for Standardisation

E	Young's modulus
EBSD	Electron Backscatter Diffraction
EDX	Energy Dispersive X-ray Diffraction
ETD	Everhart-Thornley-Detector
FGM	Functionally graded material
FIB	Focused ion beam
F_Q	Critical load
$G_{IC}(x)$	Critical energy release rate of the FGM,
G_{IC}^{ceram}	Critical energy release rates of the ceramic
G_{IC}^{metal}	Critical energy release rates of the metal
GIS	Gas Injection System
H	Magnetic field strength
H_c	Coercivity / coercive field
ICE	Ion conversion and electron detector
ICP	Inductively coupled plasma
IPF	Inverse pole figure map
IQ	Image quality map
I_y	moment of inertia
K	Stress intensity factor
K_{IC}	Fracture toughness

KLST	Single-edge bend sub-size specimen (Kleinstprobe)
K_Q	Preliminary stress intensity factor
L	Length from beginning of cantilever to loading point
LEFM	Linear elastic fracture mechanics
m_0	Slope at the fixed end due to unit bending moment
MEMS	Micro-electro-mechanical system
p	Pressure
P	Load
R	Resistivity
r_{base}	Radius of base of pillar
r_{ind}	Radius of indenter
r_{top}	Radius of top of pillar
rpm	Rotations per minute
SEM	Scanning Electron Microscopy
SEVNB	Single etch v-notched beam method
SMC	Soft Magnetic Composite
STEM	Scanning Transmission Electron Microscopy
TEM	Transmission Electron Microscopy
TRS	Transverse Rupture Strength
$V_m(x)$	Volume fraction of the metal

W	Height of the specimen for the four point bending test
x	Length from beginning of cantilever to boundary
Δx_{ind}	Deformation of indenter
Δx_{pillar}	Deformation of pillar
Δx_{sub}	Deformation of substrate
Δx_{total}	Indentation depth
z	FIB cutting depth / deposition height
α	Relative crack length
ε	Strain
σ	Stress
σ_c	Critical stress
σ_y	Yield strength
$f\left(\frac{a}{W}\right)$	Geometry factor for calculating the fracture toughness
ν	Poisson's ratio
μ_{max}	Permeability

List of Publications

The following papers have been included in this work:

Schwark, T., M. Müller, Y. Mine, T. Kreuter, O. Kraft, and R. Schwaiger, *Preparing Soft Magnetic Composites for Structural and Micromechanical Investigations*. Practical Metallography, 2017. **54**(6): p. 366-387.

Schwark, T., O. Kraft, and R. Schwaiger, *The boundaries of soft magnetic composites reveal their complexity in compression and bending tests at the micro-scale*. Materials Science and Engineering: A, 2017. **684**: p. 270-274.

Contribution to the papers included:

For above papers, I have been the main contributor to experimental work and manuscript writing. Additional experimental work has been done by Marcus Müller (ion beam polishing), Yoji Mine (EBSD and microtensile testing) and Thomas Kreuter (SEM micrographs of deformation layers) and has also been mentioned at the appropriate places in this work. Paragraphs highlighted in *italics* were taken from the above mentioned papers.

Introduction

In this research work, the deformation and fracture behaviour of the Soft Magnetic Composite (SMC) Somaloy 700 3P is analysed. SMCs consist of iron particles that are coated with an organic or inorganic insulating layer leading to low eddy current losses. The unique magnetic properties which have been extensively investigated in several studies lead to new design options with 3D flux paths. These new designs require a certain mechanical robustness of the components. The mechanical properties of the SMCs, hoewever, have not been studied as thoroughly as their magnetic properties.

In this work, the mechanical properties were studied at different length scales. The methods used to characterise the material and to analyse the mechanical properties of Somaloy 700 3P on the micro- and macroscale will be described in chapter 3. The sample preparation for micromechanical testing on Somaloy 700 3P will be explained in chapter 4. Chapter 5 will cover the material characterisation of the SMC and chapter 6 and 7 will summarise the results from mechanical testing procedures on the micro- and macroscale.

The material characterisation and the deformation and fracture behaviour of Somaloy 700 3P will be discussed in chapter 8. The challenges of the used micromechanical testing methods and the reliability of the tested mechanical properties will be evaluated in detail.

1 Background and State of the Art

With the new developments in electric mobility, optimising the energy resources of electric vehicles by using highly efficient materials is a growing market. One main objective is to increase the performance density of the electric motors. This can be realised by improving and developing new machine designs [1]. Another option is to improve the properties of used soft magnetic materials or even develop new materials. With over 95 % share of soft magnetic materials, electric sheets are the most popular material [2]. However, the sheets are limited to 2D-sample geometries. Soft magnetic composites (SMCs) with their unique magnetic properties open up a completely new field of applications [3]. They consist of pure iron particles coated with an insulating layer which gives them a very high specific electric resistance and reduces the eddy current losses. The total losses are also lower for a specific frequency range in comparison to electric sheets. The hysteresis losses are higher than for electric sheets due to the small amount of ferromagnetic iron and the remaining stresses after compression. For electric machine designing, the permeability of the material is also of importance. SMCs show a low permeability which makes them suitable for a large range of applications. In addition, the combination of their isotropic nature and new shaping options opens up the possibility to design new 3D components [4]. Those new designs, however, may demand a certain mechanical robustness of the material besides the magnetic properties of the SMC. In terms of mechanical properties, the combination of soft iron particles with a brittle insulating layer causes a rather poor mechanical behaviour of the SMC.

1.1 Soft Magnetic Materials

Magnetic materials are important for nearly every area of modern life such as electronics, computer and telecommunication [3]. **Soft magnetic materials** can be easily magnetised and demagnetised. The most important characteristics desired for a soft magnetic material are high magnetic saturation B_s, high permeability, low coercivity H_c and low core losses, all of which are depicted in Figure 1.1 [5]. The correlation between the magnetic flux density B and the magnetic field strength H is described by the hysteresis curve or B-H curve. The permeability is given by $\mu = B/H$. When the magnetic field strength is increased gradually, the domains inside the material exposed to the field get aligned gradually up to the point of magnetic saturation B_s when the material is fully magnetized. Decreasing the magnetic field strength causes hysteresis and when the field force equals zero, there is a residual magnetic induction called remanencea B_r. The coercive field or coercivity H_c is needed to bring the flux density to zero when applying the magnetic field strength in the opposite direction. By increasing the field force again, the magnetic saturation is reached without following the original curve, and the hysteresis loop is closed.

The properties of soft magnetic materials depend on chemical composition and microstructure [3]. The saturation induction rate depends on the chemical composition. Coercivity, remanence and permeability depend on the microstructure of the soft magnetic material. As cold deformation, heat treatment, grain sizes and impurities strongly influence the microstructure, the magnetic properties are influenced in the same way [6]. There are interactions of the microstructure and stresses (dislocations, grain boundaries, segregations, etc.) with the magnetic structure (domains and Bloch walls) [7].

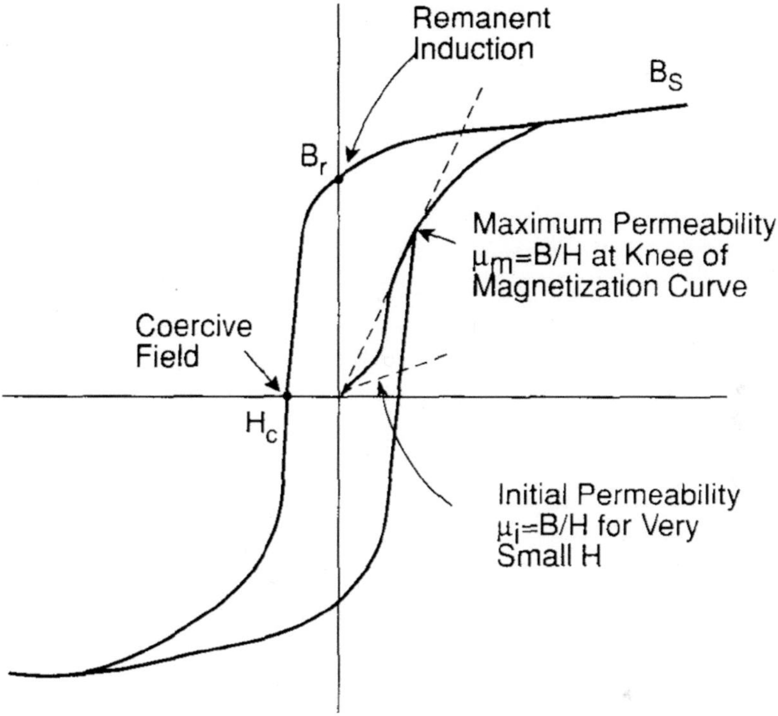

Figure 1.1: B-H curve of a soft magnetic material [5].

Core losses are a result of an alternating magnetic field in a core material. The core loss density is a function of total flux swing and the operating frequency. Hysteresis, eddy current and residual losses contribute to the core loss [8]. Hysteresis loss and eddy current losses depend on the frequency. The hysteresis loss can be determined from the area of the hysteresis loop. The eddy current losses are inversely proportional to the electrical resistivity of the material and vary with the eddy current path length (powder particle size in powder cores) [6].

1.2 Soft Magnetic Composites

Pure iron and its alloys (e.g. iron-silicon alloys, iron-aluminum-silicon alloys and nickel-iron alloys) as well as soft and hard ferrites (e.g. manganese–zinc and barium ferrites) were mainly used as laminated sheets in cores of electric machines. In order to improve the magnetic properties of these machines, not only new designs were developed but processing, properties and the effect of additives of the used materials were also investigated. Besides the improvements with traditional materials, new materials such as amorphous and nanocrystalline materials and soft magnetic composites were developed. Detailed overviews are given in [3, 9]. The most common characteristics taken into account during a design process for a core for an industrial electrical machine are: magnetic saturation, coercivity, permeability, iron losses (magnetic losses), magnetostriction (noise), delivery (forms, fully processed, semi-processed), market availability (supply, including geopolitical issues) and price [9]. Considering the first two characteristics, Figure 1.2 shows the magnetic saturation flux density versus magnetic coercivity for a wide range of soft magnetic materials including crystalline, amorphous and nanocrystalline materials as well as composites [9]. The magnetic saturation determines the possible power density of electrical machines whereas the coercivity determines the hysteresis behaviour. The presented materials cover a wide range of magnetic properties allowing applications in many fields.

Soft Magnetic Composites consist of ferromagnetic particles that are coated with an insulator (see Figure 1.3) [10]. There are different variations of particles and insulators that are powder metallurgically processed to a complex composite material. Pure Fe powder, iron-alloys with e.g. Ni, Co, Si, as well as amorphous and nanocrystalline materials, e.g. Vitroperm, are common materials for the powder particles [4, 11]. Coatings of the particles are generally divided into two categories: organic and inorganic coatings [3]. Inorganic coatings are e.g. zinc, iron or manganese phosphates, metallic oxides, and sulphates. Organic coatings are divided into thermoplastic or thermosetting resin coatings. There are also hybrid inorganic-organic coatings [12-14]. A common commercially available SMC is the Somaloy family which is described in section 1.2.3.

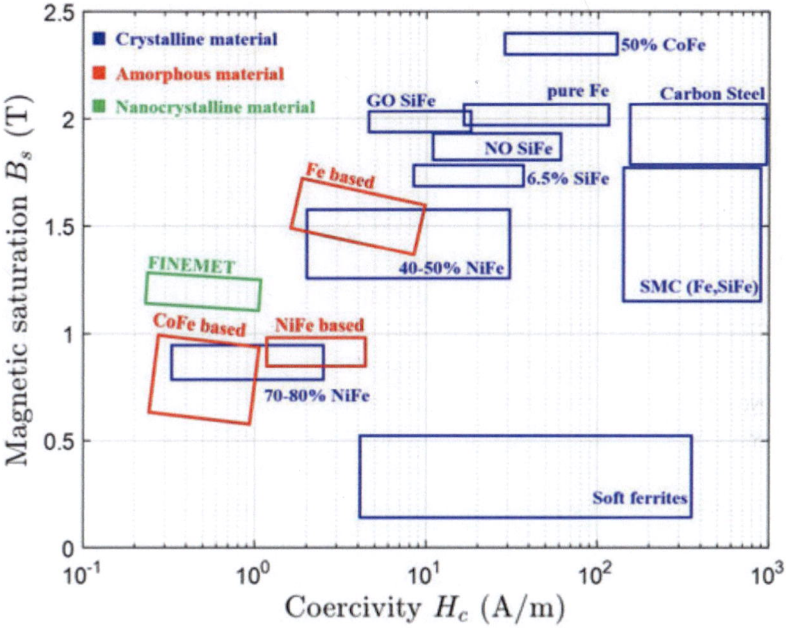

Figure 1.2: Magnetic saturation flux density versus magnetic coercivity for a wide range of soft magnetic materials including crystalline, amorphous and nanocrystalline materials as well as composites [9].

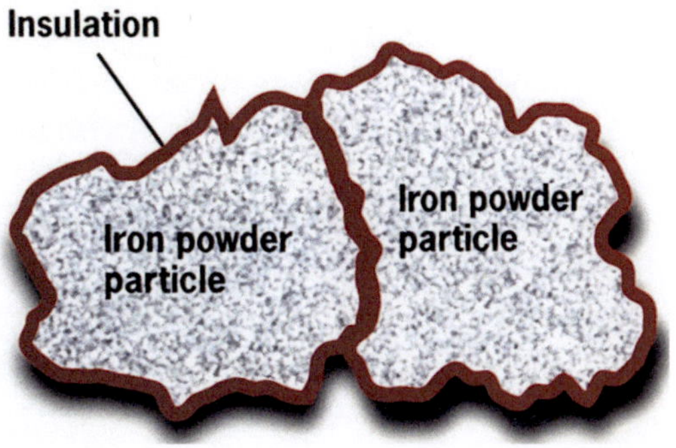

Figure 1.3: Schematic diagram of a SMC with iron powder particles and an insulating coating [10].

1.2.1 Properties

The unique electromagnetic properties of the SMCs – high resistivity and thus low eddy current loss following a relatively low total core loss at low to high frequencies – make them suitable for high frequency applications. They are also magnetically and thermally isotropic, have a large anisotropy constant, a high remanent magnetization, a high eddy current temperature and a low coercivity [3]. They also show high magnetic permeability which is, however, lower than in electrical sheets because of the porous microstructure of the SMCs [15, 16]. A further advantage of the SMCs compared to the laminated sheets is the possibility of 3D flux paths. The powder metallurgical processing of the SMCs leads, on the one hand, to a reduction in weight and size, but on the other hand, to limited sizes and shapes for cores [3, 16]. The particles of the SMCs are usually strongly cold-worked during compaction. In [17], the stress state of the particles at different annealing temperatures was investigated for Somaloy 500 and in [18], the general effect of a thermal treatment on insulated powder systems was analysed. Both studies, however, come to the general statement that increasing the annealing temperature leads to increasing recovery in the particles. Due to the powder metallurgical processing of the SMCs, their mechanical strength is generally lower than that of laminated magnetic materials [19-21]. The electromagnetic and mechanical properties depend on not only the chosen powder but also on the added lubricant or binder and the compaction process [16]. The additives determine the maximum temperature for the heat treatment. The SMCs are not sintered because the dynamic losses would increase rapidly with frequency [16]. Increasing the heat treatment temperature and reducing the amount of additives leads to improved magnetic properties but also to a lower strength of the SMCs [16]. The Ashby-Diagram in Figure 1.4 shows the tensile strength versus the fracture toughness of different traditional soft magnetic materials and SMC. While all other materials cover a wide range of properties, the SMCs are in the lower left corner providing a huge potential for improvement considering their mechanical properties.

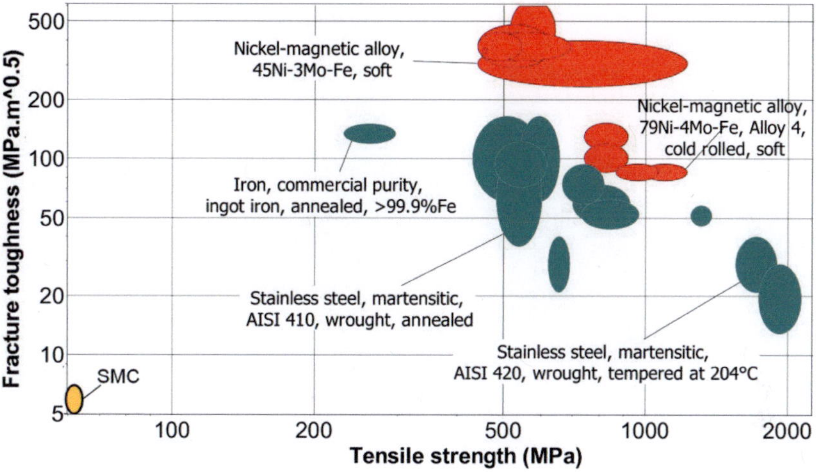

Figure 1.4: Ashby Diagram showing fracture toughness vs. tensile strength of different traditional soft magnetic materials and SMC [22].

1.2.2 Applications

In general, the applications of soft magnetic materials can be divided into alternating current (AC) and direct current (DC) applications as well as low and high frequency applications. For DC applications, permeability, coercive force and saturation induction are the most relevant magnetic characteristics. For example, a brushless DC motor can be realised [16, 23] (see Figure 1.5a). For AC applications, parameters such as permeability, saturation and total core loss are the key magnetic characteristics. In particular for AC applications, the possibility to have a 3D-flux path can be used for new machine designs, e.g. flux machines, tubular flux machines and claw pole rotor machines [24] (see Figure 1.5b). For the claw pole rotor machines and also for axial flux motors, the application of SMCs can reduce the eddy current losses [25].

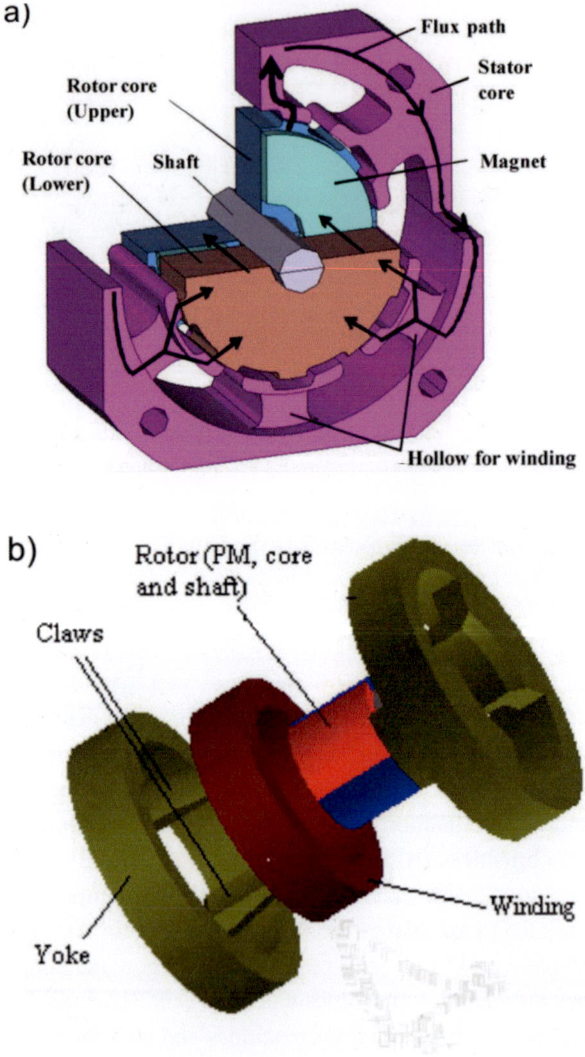

Figure 1.5: a) Brushless DC motor [23], b) claw pole motor [24].

1.2.3 Somaloy

The Somaloy family are commercially available SMCs from Höganäs AB, Sweden. They all consist of water-atomised pure iron particles with an inorganic, phosphorous coating but differ in their heat treatments leading to different magnetic and mechanical properties. The iron particles are first pressed and then heat-treated in air, water or nitrogen. The selected atmosphere results in different particle boundaries [16, 26]. Table 1.1 summarises magnetic and mechanical properties of different members of the Somaloy family. The high resistivity of Somaloy 700 1P and 700 3P indicates low eddy current losses. Varying the compaction pressure and atmosphere results in a higher transverse rupture strength (TRS) of 125 and 140 MPa of Somaloy 700 3P and 1000 3P, respectively, compared to the low 40 and 50 MPa of Somaloy 700 1P and Somaloy 500 1P, respectively. The magnetic flux density as well as the permeability increase from Somaloy 500 1P to 1000 3P. [26] In addition, the pure iron particles are ductile, whereas the phosphorous coating is brittle. This combination typically results in rather poor mechanical behaviour.

Somaloy 500 1P contains a binder and has a mean particle size of about 100 μm [27]. It is generally pressed at 800 MPa and then heat treated at 500°C in air. Somaloy 500 1P contains a phosphorous layer of ~20-30 nm thickness [28, 29]. The macroscopic mechanical properties have already been investigated in several studies. With a TRS of 40-50 MPa, the strength of this material is quite low compared to other core materials [26, 30]. The microhardness of the material shows an increasing hardness at increasing densities with a maximum of 80-100 HV. Nanoindentation of individual particles gave a hardness of about 1.5 GPa - 2 GPa. [17, 31] The fracture toughness of the material is about 1.8 - 2.1 MPa√m [32].

Table 1.1: Magnetic and mechanical properties of the Somaloy SMCs and their applications [26].

Material	Somaloy 500 1P	Somaloy 700 1P	Somaloy 700 3P	Somaloy 1000 3P
Resistivity [μΩm]	70	400	200	70
Transverse Rupture Strength TRS [MPa]	50	40	125	140
Magnetic flux density B at 4 kA/m [T]	1.26	1.31	1.37	1.42
Magnetic flux density B at 10 kA/m [T]	1.51	1.56.	1.61	1.63
Permeability μmax	500	540	750	850
Application	One of the most known SMC materials on the market and a base line for fast switching actuators.	Suitable for both fast switching actuator and ignition. A good material for geometries with high aspect ratios.	High magnetic performance and high strength makes this a very good material for fast switching actuators for fuel injectors but also for ignition cores with high demands.	Optimised for maximised B-H characteristics and strength

The mean particle size of Somaloy 700 3P is larger than that of Somaloy 500 [12]. There have been several studies analysing the influence of the particle sizes on the mechanical and magnetic properties [33, 34]. Somaloy 700 3P is generally pressed uniaxially at a pressure of 800 MPa. The compacted specimen is annealed at 530°C under water vapour. The water vapour causes the formation of an additional iron oxide layer between the particles. Figure 1.6 shows a model of the complex boundary layers of Somaloy 700 HR-3P. There is an iron oxide layer at the particle interface and a phosphorous layer in the middle. The phosphorous layer can also appear as a thin layer between two iron oxide layers. At different annealing temperatures, the phosphorous layer is either amorphous (530 °C) or crystalline (700 °C) [33].

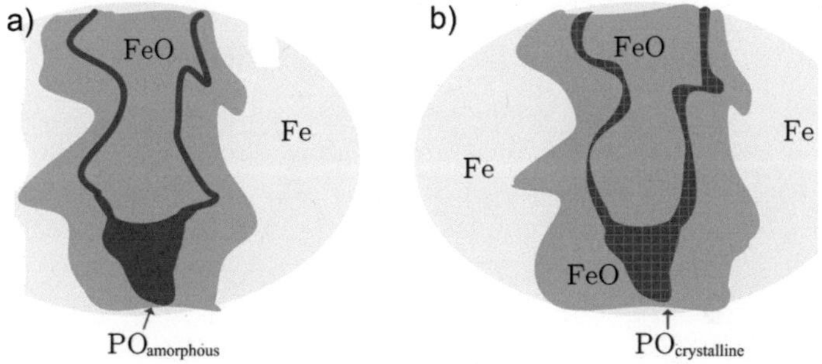

Figure 1.6: Model of the particle boundary of Somaloy 700HR- 3P [33]. a) An annealing temperature of 530°C results in an amorphous phosphorous layer which can be either in the middle of an iron oxide layer or between two iron oxide layers. b) An annealing temperature of 700 °C results in a crystalline phosphorous layer.

The chemical composition of the particles and boundaries of Somaloy 700 HR-3P was analysed by Scanning Transmission Electron Microscope (STEM) and Energy Dispersive X-Ray (EDX) spectroscopy. Figure 1.7a shows an STEM micrograph of the complex boundary of Somaloy 700HR-3P annealed at 530°C [33]. The EDX line scan in Figure 1.7b shows that the iron particle itself is pure iron, the first layer contains iron oxide and the middle layer contains phosphorous. The additional iron oxide layers of Somaloy 700 3P were shown to improve the overall mechanical properties [12, 26, 35].

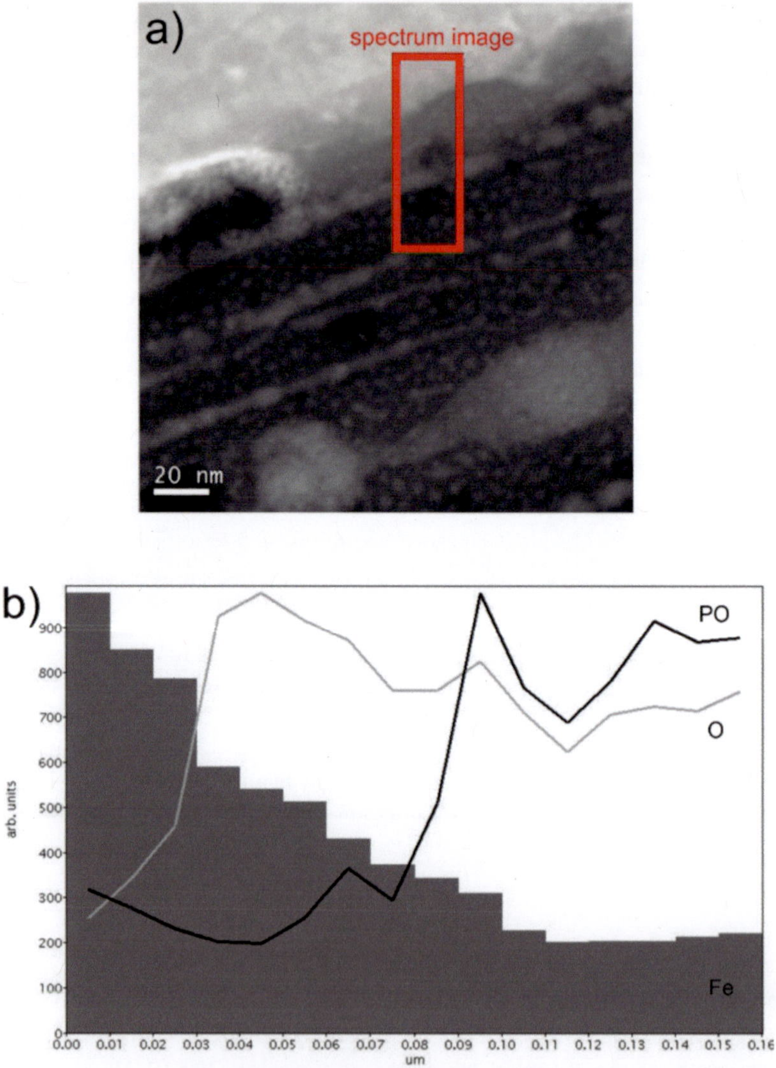

Figure 1.7: a) STEM image of a Somaloy 700HR-3P boundary annealed at 530°C, b) EDX line profile of the boundary [33].

Figure 1.8 shows an Electron Backscatter Diffraction (EBSD) image of Somaloy 700HR-3P. The boundaries could not be resolved but there is a clear misorientation in the particles. The iron particles are strongly cold-worked during compression. The heat treatment at 530°C did not lead to a full recovery which is shown by the remaining misorientation in the EBSD scan [33].

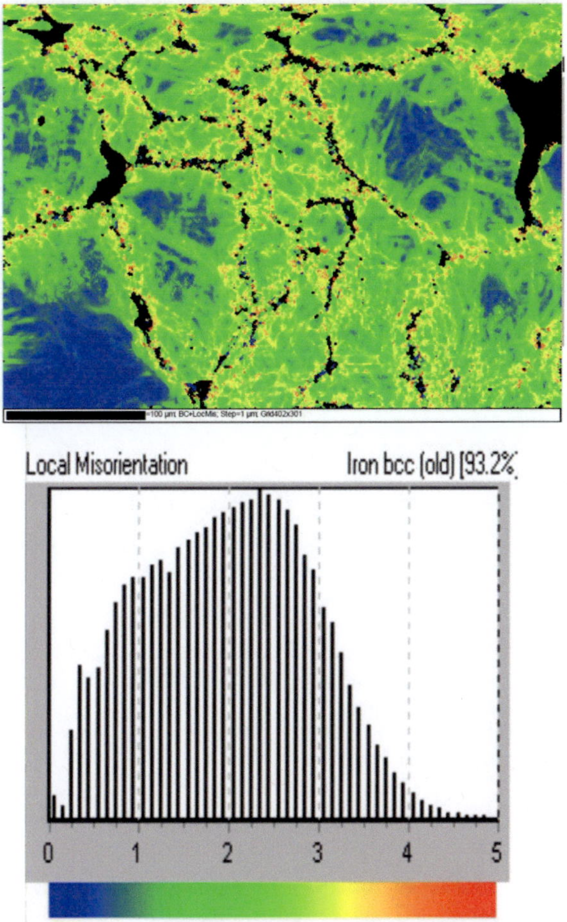

Figure 1.8: EBSD image of Somaloy 700HR-3P with clear misorientations in the particles. The boundary layer could not be resolved [33].

At the macroscale, some mechanical properties of Somaloy 700 3P were investigated. Figure 1.9 shows the strong tension-compression asymmetry of Somaloy 700 3P. The density of the material influences the compressive strength but not the tensile strength. The compressive strength is more than three times higher than the tensile strength. [35] Other studies show similar values for the TRS of 29-30MPa and the yield strength of 264-279 MPa [34]. The low compressive strength indicates a brittle behaviour of the material whereas the SMCs show a rather deformable behaviour under compression. The Brinell Hardness of Somaloy 700 3P is 91-107 HB [34], indicating a soft material.

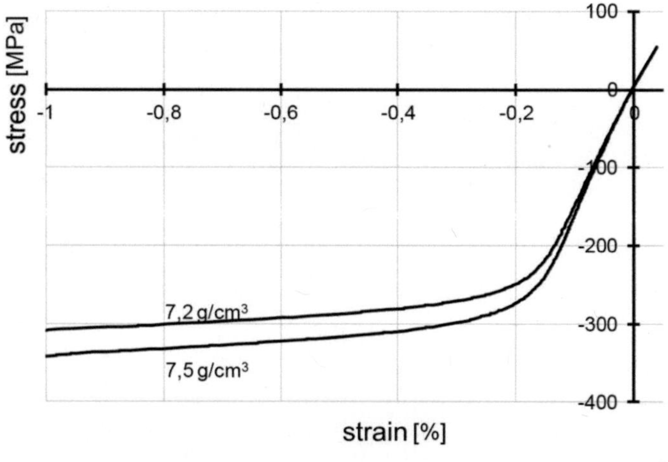

Figure 1.9: Stress-strain diagram for tension and compression of Somaloy 700 3P. There is a clear tension-compression asymmetry. The density of the material influences only the compressive strength [35].

1.3 Micromechanical Testing Methods

In the fast growing market of miniaturised technological systems, such as micro-electro-mechanical systems (MEMS) or medical and microelectronic devices, the structural elements are more and more in the range of a few tens of nanometres up to a few hundreds of microns. The dimensions of conventional mechanical testing are too large to test these small scale devices and novel techniques are required to determine the properties at this scale. Not only in terms of industrial applications but also in the bio-medical area, small scale testing methods are in high demand. In addition, there is a resulting demand for a better fundamental understanding of materials properties in small dimensions. "Smaller is stronger" is one of the phenomena that has been investigated [36]. Metallic thin films, for example, show higher strengths compared to their bulk counterparts because the dislocations are restricted within the film in presence of the substrate [37]. *Nanoindentation has become a powerful method for small-scale testing [38-40]. In addition, new test geometries have been developed exploiting the high-resolution force and displacement measurement capabilities of the nanoindenter, thereby allowing the investigation of the mechanical properties in compression, tension, bending, shear as well as fatigue and fracture toughness testing at the micro and nanoscales.* [41]

1.3.1 Nanoindentation

One of the most frequently used techniques for small-scale testing of mechanical properties is nanoindentation. There are many advantages to use this technique: it can be applied to a large range of materials, it only requires a smooth surface, which makes sample preparation in principle easy, it is an easy and straightforward technique, and it is considered to be non-destructive as the indentation depths are in the nanometre and micron range.

The principal components of a nanoindenter are the indenter tip, the actuators and sensors, respectively [42]. When pressing the indenter tip into the surface of the tested materials, the mechanical load and displacement are applied and recorded by the sensors and actuators. Different commercially available setups are described in [43]. Recently, a number of setups for in-situ testing inside the Scanning Electron Microscope (SEM) or Transmission Electron Microscope (TEM) have been developed [44, 45].

Nanoindentation testing can be performed quasi-statically or dynamically. In quasi-static testing, the load is applied, held for a certain amount of time and then removed. The contact stiffness, which is needed to determine modulus and hardness of the material, is then determined from the unloading data. In dynamic testing, a small amplitude force oscillation is superimposed at a relatively high frequency onto the quasi-static load signal, and the displacement response is measured [43, 46]. This method allows the continuous determination of the contact stiffness during loading.

From the typical load-displacement curve as shown in Figure 1.10 material characteristics such as hardness and Young's modulus can be obtained by using the slope of the elastic unloading region (dP/dh). The detailed analysis in described in [47].

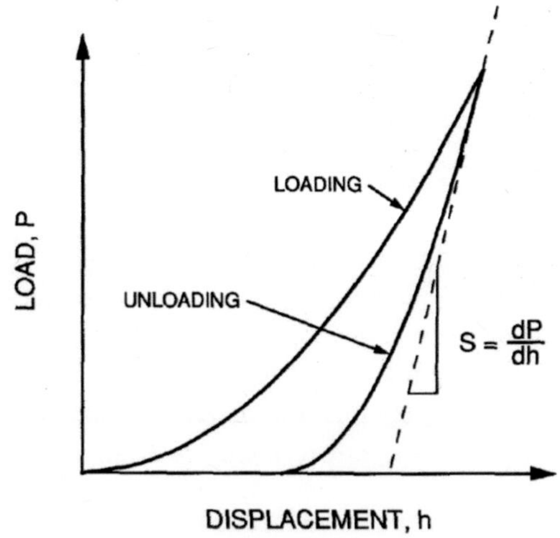

Figure 1.10: Typical load-displacement curve [47].

The indenter tip should be made of a material with at least 20% higher hardness than the tested material [48]. Diamond as the hardest material is commonly used for indenter tips. However, at elevated temperatures, other materials such as boron nitride are preferred [48]. There are many shapes of indenters and their applications depend on the tested material and the mechanical properties to be determined. [49] The most commonly used tip is a Berkovich tip with a triangular pyramidal shape. Cube Corner and Vickers tips are used to determine the fracture toughness. Spherical tips are used to determine yielding and work hardening. Flat punches and conical flat punches are preferably used for polymers and bio-materials to determine storage modulus and loss modulus [50].

Not only Young's modulus and hardness can be determined but also several other material characteristics such as yield strength, creep behaviour or fracture toughness [51, 52]. Scratch or wear tests can also be performed [52-54]. Furthermore, the nanoindenter can even be used as a scanning probe to determine surface topography or can be combined with an atomic force microscope [55, 56]. With suitable heating and cooling stages, temperature-dependent testing is also possible [57].

1.3.2 Microcompression Tests

Besides nanoindentation, microcompression testing is also a popular method for small scale testing. Figure 1.11 shows the setup of a microcompression test. A pillar made from the bulk material is compressed by the flattened indenter. The indenter tips can be a Berkovich or a cone tip that are truncated [58]. The advantage compared to the indentation methodology is the relatively uniform and uniaxial deformation [59]. The tests can be carried out ex-situ in a nanoindenter or in-situ in an SEM. The advantage of in-situ experiments is that the evolving mechanisms during compression can be observed. Ex-situ experiments, though, are easier to perform.

The micropillars can be produced by different methods e.g. focussed ion beam (FIB)-milling, inductively coupled plasma (ICP) etching or electroplating. FIB-milling is a commonly used technique to fabricate small-scale specimens with a focussed Gallium ion beam. There is, however, a vivid discussion about the influence of Gallium contamination on the mechanical properties of the specimens. [59, 60] The top surface of the compressed volume can be circular, square or rectangular, and sizes range from a couple hundred nanometres to tens of microns in diameter. The fabrication time also varies between hours and days depending on size and material. There are two methods for fabricating pillars that can be differentiated by the direction of the ion beam to the sample surface. In the first method, the orientation of the ion beam is perpendicular to the sample surface. The pillars are easy to fabricate and can have square shaped or circular top surfaces. However, the pillars show some taper of about 2° to 5°, meaning that the top area is smaller than the bottom area. There are methods to avoid or reduce the taper by tilting the sample by several degrees, as done in the lathe milling in which the ion beam is at angle to the sample surface [61]. It is a time-consuming method but it leads to a controlled aspect ratio and diameter of the micropillars.

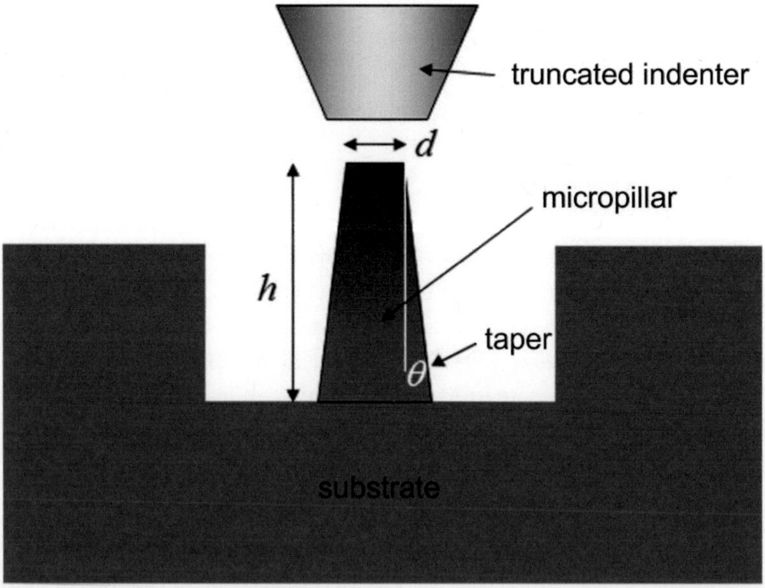

Figure 1.11: Schematic of a micropillar compression test adapted from [62].

As the microcompression test is not standardised, there are several aspects that have to be considered: i) the aspect ratio of the micropillars (the ratio of height to diameter of the micropillar), ii) the influence of the substrate / material below the pillar as the substrate has a certain stiffness and, thus, there will be a certain sink in of the pillar during compression, iii) the taper angle of the micropillar which leads to a non-uniform applied stress field within the pillar, iv) the missing control of the friction between the indenter and the micropillar and v) the misalignment between the compression direction and the micropillar axis. [59, 63]

The deformation mechanisms at all length scales as well as extrinsic and intrinsic size effects have been intensively studied with this microcompression testing. First, single-crystalline materials were used to understand fundamental mechanisms, but soon also more complex materials were used such as isotropic, anisotropic and nano-structured materials. Microcompression has also been used on MEMS structures as their testing volume is too small for conventional compression tests. [59, 63]

1.3.3 Microcantilever Bending Tests

The nanoindenter cannot only be used for indentation and compression testing but also for bending testing. A Berkovich or wedge tip is used to apply a point or line load on a microcantilever. The tip mechanically bends the cantilever while load and deflection are monitored. Not only Young's modulus and yield strength can be obtained but also fracture toughness if a crack is induced in the cantilever [64, 65]. As for the other small-scale tests, there is also the possibility for in-situ testing within an SEM [66].

The fabrication of microcantilevers requires more effort compared to micropillars. The cantilevers can be etched but the dimensional control is difficult. Another possibility is to use FIB-milling. Cantilevers can be fabricated either at the edge or in the bulk of the sample. Fabricating at the edge of a sample has the advantage that cantilevers with a rectangular cross section can be easily processed. However, it requires a very smooth sample edge (see Figure 1.12a) [67]. In contrast to cantilevers with a rectangular cross section, cantilevers with a pentagonal or triangular cross section can be placed at any position on the surface making this method suitable for e.g. anisotropy studies [68]. These cantilevers are produced by tilting the sample and, thus, removing the material underneath the cantilever (see Figure 1.12b and c) [68, 69]. Single crystalline materials and in particular single grains, selected grains or grain boundaries of polycrystalline materials have been investigated [68].

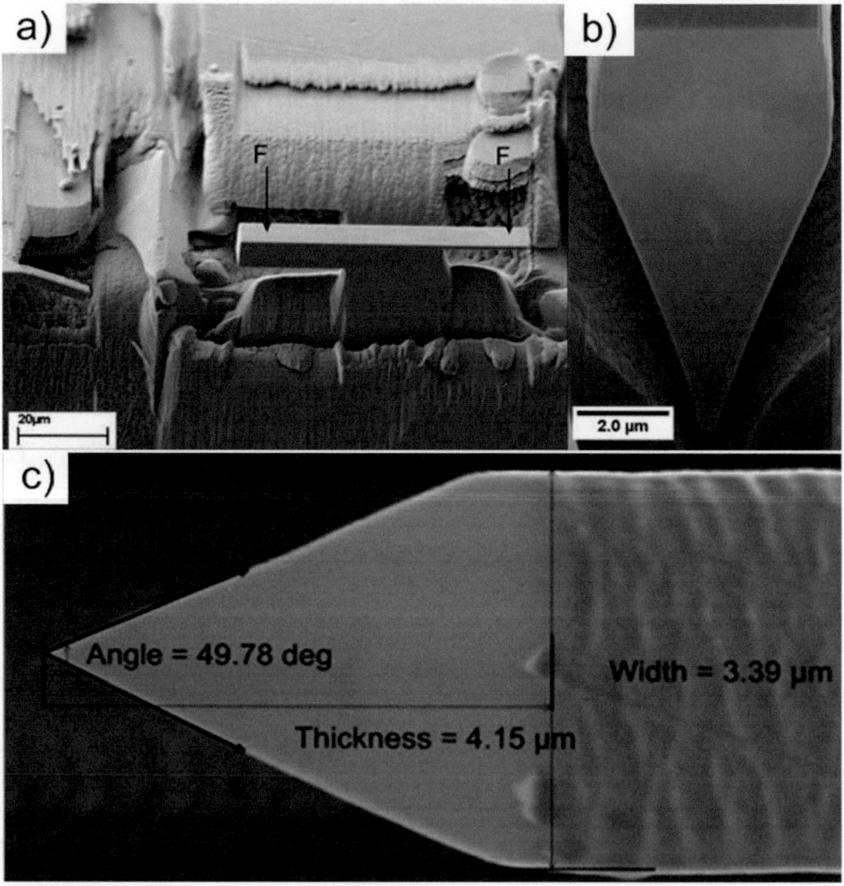

Figure 1.12: Different types of microcantilevers : a) Cantilevers with a rectangular cross section at the edge of a sample [67], b) Cantilever with a pentagonal cross section [69] and c) Cantilever with a triangular cross section [68].

1.3.4 Microtensile Tests

In microtensile tests, the load is applied uniaxially and the resulting deformation is nominally uniform, which makes data interpretation straightforward. [70] Specimen sizes can vary from several millimetres to tens of nanometres.

For fabricating tensile test specimens, there are different possible preparation routes depending on the size of the sample. For larger scale sample, the preparation routes are presented in [71]. Many tensile test samples on the micro or nanoscale need support structures whereas free-standing specimens, e.g. nanowires or whiskers, are either processed by e.g. etching or FIB-milling [72]. Support structures can be prepared by microfabrication or MEMS processing with both additive and subtractive methods [73]. Another method is to use thin film materials [74]. Thin platelets were cut out from the tested material and were thinned down to approximately 20 μm by mechanical polishing and ion beam polishing. Next, the gauge section was prepared by FIB-machining (see Figure 1.13b).

The tests can be performed ex-situ or in-situ in an SEM or a TEM. There are different setups for ex-situ testing depending on the sample size. For samples in the millimetre range, the samples can be gripped, whereas for smaller scale samples they may be glued. A setup with grips in which the sample can self-align was developed in [75]. There are also systems directly using On-Chip tensile testing setups [76]. Nano-sized materials like gold nano-wires can be tested in-situ in an SEM or a TEM [77-79]. A micro-gluing grip uses instant glue to fix the micro-sized specimens [74, 80-83]. The tensile load could be applied by piezoelectric actuators and measured by a load cell. The elongation was measured by gauge mark displacement. The micro-sized specimens were fixed to the testing system by an instant glue (see Figure 1.13) [70].

Applications of tensile tests are versatile. Biological materials can be analysed to investigate the mechanical properties of the hierarchical structures of those materials [84]. Other examples are the investigation of the size-dependent behaviour in single-crystalline copper [85] or the determination of deformation mechanisms in single-crystalline metals [72].

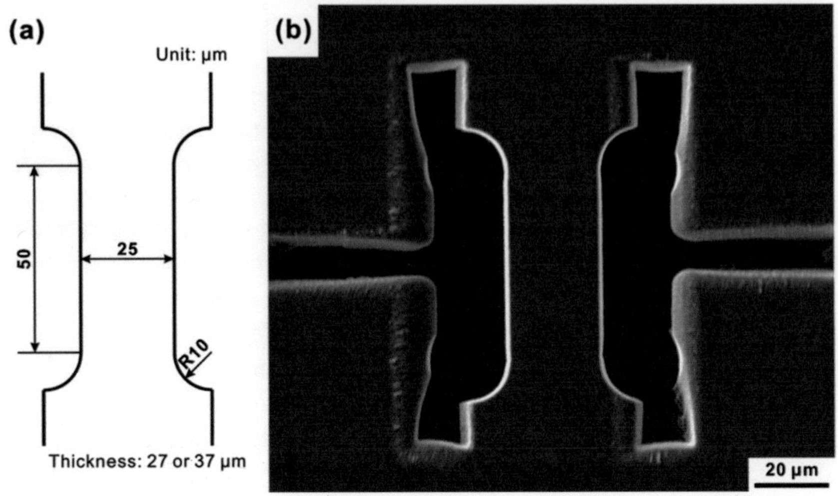

Figure 1.13: Microtensile test sample for a micro-glue tensile test [80]: a) Dimensions of the specimen, b) SEM image of the FIB-milled specimen.

1.4 Theory and Methods of Fracture Toughness Testing at Different Scales

Real materials are never defect-free. Cracks are particularly important defects as they cause stress concentrations in the material and can lead to an uncontrolled failure of the device. The material deformation behaviour can be described as being linear elastic, nonlinear elastic, or elastic-plastic. The linear elastic fracture mechanics (LEFM) assumes a purely linear elastic deformation behaviour and an ideally sharp crack tip with an infinitely high stress concentration at the tip. Irwin introduced a K-factor to describe the intensity of an elastic crack-tip field [86]. In real materials, however, a plastic zone forms at the crack tip and the LEFM is not valid anymore. In this case, the J-integral was proposed by Rice to characterise the intensity of the elastic-plastic crack-tip field [87]. The fracture behaviour correlated with the micro-mechanism of fracture being described as ductile or brittle. For brittle fracture, the LEFM applies and K-factors are used whereas in the case of ductile fracture, the

J-integral for elastic-plastic behaviour is used. There are also methods considering a ductile-brittle transition fracture. Compared to metals, ceramics require a different approach due to their brittle nature. In the following, the procedure for ceramics will be discussed. The stress intensity factor K is given by:

$$K = f\left(\frac{a}{W}\right) \cdot \sigma \cdot \sqrt{a} \tag{1.1}$$

where $f\left(\frac{a}{W}\right)$ is a geometry factor, σ is the stress and a is the characteristic size of a defect. Samples with different flaw sizes will lead to a scatter in the fracture strength. Thus, the thickness, size and shape of the testing sample as well as the material density and surface finish have a strong effect on the evaluation of the fracture toughness.

These effects are accounted for by using the LEFM approach as shown in equation 2:

$$\sigma_c = \frac{K_{1C}}{Y \cdot \sqrt{a_c}} \tag{1.2}$$

where σ_c is the critical stress, a_c the critical defect size and K_{IC} is the fracture toughness of the tested material. When comparing the different possible testing methods, e.g. tensile testing and three- and four-point bending, the flexural strength obtained in tension could be significantly lower than in three-point bending. Similarly, the flexural strength from a three-point bending test could be higher compared to a four-point bending test because the critical defect may not be located at the maximum bending moment. Standard terminology relating to fracture toughness testing and its evaluation has therefore been defined by the American Society for Testing and Materials (ASTM) as well as by the German Institute for Standardisation (DIN) [88, 89] describing e.g. sample geometries, testing procedures and equations. [90, 91]

At the microscale, the measurement of the fracture toughness is difficult due to the size of the samples. Particularly when testing thin films, a simple downscaling of macroscopic test methods is not easy to implement. However, there are several methods to determine the fracture toughness [52]. A more classical approach is to use nanoindentation with a sharp pyramidal indenter. Another method is to use simple compression experiments with test structures of a specific geometry, i.e. a pre-notched double cantilever [92]. In the splitting pillar experiment, there is no need to measure the crack length or fracture surfaces, which makes it attractive for fracture toughness measurements of thin film and small-scale devices [93]. A classical cantilever with a predefined notch can also be used [64, 65]. For nanoindentation, smooth surfaces are needed. Pillars, double-cantilevers and cantilevers are commonly milled using a FIB system. Detailed information about the fabrication of the test structures is given in the previous sections. Not only thin films are tested but the fracture toughness of biomaterials, such as dental enamel, can also be determined [65]. Another example is to measure the fracture toughness of individual grain boundaries in bismuth embrittled copper with pre-notched triangular shaped cantilevers [69].

2 Objectives

In this research work, there are two main objectives: i) the mechanical characterisation of the SMC and ii) to apply micromechanical testing methodologies to a complex, technologically relevant material. The SMC used in this work is a commercially available Somaloy 700 3P. It consists of pure iron particles that are coated with an insulating phosphorous layer. The powder is compressed at 800 MPa and annealed at 530 °C under water-vapour. The water vapour causes a strength-enhancing additional iron oxide layer which has not been investigated in detail. The aim of this research work is to understand the importance of the layers and to find an explanation for the overall improved mechanical properties of the Somaloy 700 3P. For this, the boundary has to be characterised in detail by means of LM, SEM, TEM, EBSD and EDX as well as quantitative microstructural analysis. The mechanical properties of the boundaries as well as the particles are determined by micromechanical testing methods i.e. nanoindentation, microcompression, microcantilever bending and microtensile testing. To complete the material analysis, macroscopic four-point bending tests were also performed to determine the fracture behaviour. The influence of the boundaries on the macroscopic and microscopic mechanical behaviour was investigated. The mentioned micromechanical testing methods are often applied to model materials with defined properties, i.e. defined crystal orientations, single-crystalline state or a defined grain boundary. Applying them now to a commercially available complex composite material leads to the question of sample preparation, practicability, reliability and limits of the test methods.

3 Experimental Details

3.1 Somaloy 3P 700

Somaloy is a commercial powder metal from Höganäs AB, Sweden. The particles are formed by water-atomisation: a stream of molten pure iron enters an atmosphere resulting in the formation of small iron particles. The particle shape is determined by the impact media, i.e. in the case of Somaloy, water – giving its particles their irregular shape. The iron particles are then coated chemically with an inorganic, phosphorous layer (Somaloy 500) and in case of Somaloy 700 an additional binder layer. [94] Somaloy 700 3P is a so called "mesh 40" powder indicating the particle size of below 425 µm. The particle size distribution is given in Table 3.1 [12]. From a mechanical point of view, the pure iron is soft, whereas the phosphorous layer is brittle. This results in a combination of soft and brittle material characteristics.

Table 3.1: Particle size distribution of Somaloy 700 3P [12]

Particle Size [µm]	Percentage [%]
0 – 75	3
75 – 150	29
150 – 212	29
212 – 425	39
425	0

The samples used in this work were all made from Somaloy 700 3P. The samples were fabricated at Höganäs AB, Sweden, as prismatic LTRS (Long TRS) bars of dimensions 90 mm x 12 mm x 3 mm with rounded corners with a radius of about 2 mm (see Figure 3.1a). They were uniaxially pressed at 800 MPa up to a density of 7.6 g/cm^3 and then annealed at 530 °C in water vapour. The vapour treatment led to additional iron oxide layers in the material. Somaloy

500 and 700 1P did not form these additional layers because they were annealed in air resulting in a lower TRS (see Table 3.2) [12, 26].

Table 3.2: Mechanical properties of Somaloy 500 1P and 700 3P for different processing parameters. [95]

	Somaloy 500 1P	Somaloy 500 1P	Somaloy 700 3P	Somaloy 700 3P
Additive	0.5% Kenolube	0.5% Kenolube	0.3% 3P Lube	0.3% 3P Lube
Compaction Pressure Die temperature	 600 MPa RT	 800 MPa RT	 600 MPa 80°C	 800 MPa 80°C
Heat treatment Atmosphere Temperature	 Air 500°C	 Air 500°C	 Steam 530°C	 Steam 530°C
Transverse Rupture Strength (TRS)	55 / 55	50 /50	130 / 130	125 / 125
Tensile Strength/ Yield strength	15 / 15	15/15	65/65	65 / 65
Compressive Strength / Yield	250 / 180	310 /220	350 / 260	550 / 280
Young's modulus	120	110	190	190

In order to vary the degree of oxidation, the thickness of the LTRS samples was increased to 9 mm (see Figure 3.1b). The density was also 7.6 g/cm^3. During annealing, the atmosphere and the temperature were kept constant, while the processing time was varied in order to obtain different oxidation gradients. For fully oxidised SMCs, the samples were kept longer inside the oven resulting into a full oxidation of the whole samples. Partially oxidised samples were kept in the oven for a shorter time and thus had a different boundary type distribution including a core region without any oxidation.

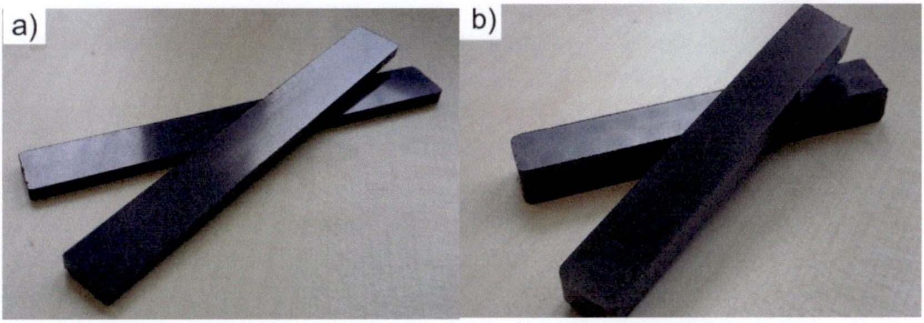

Figure 3.1: a) LTRS samples made of Somaloy 700 3P, b) TRS samples of 9 mm thickness made from Somaloy 700 3P under two different annealing conditions.

3.2 Metallographic Preparation

3.2.1 Sectioning and Mounting

Sectioning induces the most severe damage during specimen preparation. However, this damage can be minimised by using a hard abrasive wheel without any coolant. The specimens were first cleaned and then fixed on electron microscopy stubs with heat resistant glue (M-Bond 610, Vishay Precision Group). After that, the stubs were mounted directly onto a nanoindentation specimen holder which was put upright into a silicone mould. Next, resin (EpoThin 2, Buehler GmbH, Düsseldorf, Germany) was poured into the silicone mould until it completely covered the specimen and the holders. Then, the mould was put into a vacuum chamber (-0.8 bar) for 30 min (impregnation system, ATM, Mammelzen, Germany). Infiltration of the macro-pores with resin prevents any breakouts of the material and an accumulation of dirt in the pores during further preparation steps. After the resin had been cured, the mould was removed. Finally, after polishing, the resin was removed by heating the specimen to 125°C for 15 min in a heating cabinet. The resin softens at this temperature and can be removed mechanically. [41]

3.2.2 Grinding and Polishing

Table 3.3 lists all the preparation steps including the grinding papers, suspensions and cloths used. Grinding and polishing were performed on a Saphir 520 (ATM, Mammelzen, Germany). As the samples had been mounted upright, step 1 was to remove the resin still covering the surface. Additional grinding was required in order to remove the sectioning damage and to produce a planar surface. Silicon carbide (SiC) grinding papers from different suppliers were used (ATM, Mammelzen, Germany, Buehler GmbH, Düsseldorf, Germany or Cloeren Technology GmbH, Wegberg, Germany) with water as the lubricant in all grinding steps. From step 2 onwards, a specialized specimen holder (specimen holder M, Heraeus Kulzer, Wehrheim, Germany) was used to simplify the specimen handling. The recommended handling instructions distributed by the supplier, which describe fixing the specimens in the holder, did not work for our specimens. Thus, the specimens were only inserted into the holder without fixing them. After grinding (steps 2-4), the pores of the SMCs were almost completely smeared. The following difficulties needed to be handled during the next preparation steps: (i) all pores except for the macropores had to be opened, (ii) the soft and brittle components required different polishing procedures and (iii) the pure iron corroded easily. Two different cloths were used for polishing, a satin woven acetate cloth (MD-DAC, Struers) and a short nap, soft synthetic cloth (Zeta, ATM, Mammelzen, Germany), which abraded the brittle component and reduce damage from the brittle remnants in the soft component. Oil-based diamond suspensions (3 µm and 1 µm grain size, ATM, Mammelzen, Germany) and alcohol-based (0.25 µm grain size, microdiamant, Germany) prevented the corrosion of the SMC. The lubricant used was water and oil-based (DP-Lubricant Red, Struers, Erkrath, Germany). For steps 5-7, a squeeze bottle was used keeping the cloth moisturized at all times. Another approach for the final mechanical polishing step was an alumina-based fine-grained suspension (Eposal, ATM, Mammelzen, Germany) on a neoprene cloth (MD-Chem, Struers, Erkrath, Germany) (step 7b). All steps were performed at 150 rotations per minute (rpm). [41]

Table 3.3: Grinding and polishing steps with corresponding cloths, suspensions and lubricants for the mechanical polishing preparation of the SMCs [41].

	Step	Paper/Cloth	Suspension	Lubricant	rpm	Time	Remarks
Grinding	1	P240		water	200	until planar	removes moulding material
	2	P1200		water	200	until planar	from here with sample holder
	3	P2500		water	200	1-2 min	
	4	P4000		water	200	2 min	
Polishing	5	Struers MD-Dac	3 µm oil-based	oil-water based	150	3 min	repeat if corrosion occurs
	6	ATM zeta	1 µm oil-based	oil-water based	150	3 min	repeat if corrosion occurs
	7 a	ATM Zeta	0.25 µm alcohol-based	oil-water based	150	3 min	
	7 b	Struers MD Chem	Eposal (0.06 µm)	water	150	1 min	

3.3 Ion beam Polishing

Ion beam polishing was carried out in vacuum (2 x 10^{-4} mbar) using a triple ion beam cutter TIC 3X (Leica Microsystems, Vienna, Austria). We used a rotary stage for planar polishing. The specimen was moved laterally just below the cross-over point of the ion beams, while at the same time the specimen holder was rotated. In order to achieve homogeneous results, the specimen surface had to be well-aligned with the plane of rotation. Depending on the incident angle, different levels of material removal can be accomplished. Incident angles of 10°–15° promote the ablation of material, while angles of 3°–5° are used to obtain smooth surfaces. As reported in [96-100], ion beam polishing facilitates the preparation of complex materials; preparation artefacts and deformation can in principle be eliminated over a large surface area by using crossed Argon ion beams. The ion beam polishing was combined with different mechanical polishing procedures. The parameters are listed in Table 3.4. [41]

3.4 Light Microscopy

A reflecting light microscope Eclipse LV150N (Nikon, Tokyo, Japan) was used to control the individual metallographic preparation steps. Dark field microscopy was used to detect polishing scratches and possible material breakouts as well as to control the cleaning state of the samples. In dark field microscopy, the central light used in bright field microscopy is blocked [101]. A condenser forms a hollow cone with light travelling around the cone rather than through it. This way, only oblique rays of light strike the sample. The image is formed by rays scattered by the sample and captured in the objective lens. Contrast is created by bright outlines on a dark background. Edges, boundaries and scratches appear bright. In contrast, the fracture surfaces were imaged by bright field microscopy in which the central light is used for imaging.

Table 3.4: Parameters used for ion beam polishing in combination with different mechanical surface preparations. Procedures A-E describe planar polishing with a rotary stage; CS-1 and CS-2 describe cross-sectional ion beam polishing [41].

	Procedure	Final mechanical polishing step		rpm	Time	Remarks
				Time [h]	Angle [°]	Voltage [kV]
Planar ion beam polishing	A	Step 5	cleaning polishing	0,1 3	12 4,5	from here with sample holder
	B	Step 6	cleaning polishing	0,1 3	12 4,5	
	C	Step 7a	cleaning polishing	0,1 3	12 4,5	
	D	Step 7a	cleaning polishing	– 5	– 10,5	repeat if corrosion occurs
	E	Step 7a	cleaning polishing	0,5 5	12 3	repeat if corrosion occurs
Cross-sectional ion beam polishing	CS-1	P2000	polishing	6	0	
	CS-2	P2000	polishing	4	0	

3.5 Quantitative Microstructural Analysis

The point count method described in ASTM 562 [102] was used to determine the fractions of types of particle boundaries in fully and partially oxidised SMCs. Images were taken in the SEM (Helios Nanolab 650 Dual Beam, FEI Company, Hillsboro, OR, USA). Roughly 120 images were needed to cover the complete width of the specimen. The resolution of the images was chosen in a way that the types of particle boundaries can be identified. A rectangular grid of 20 x 30 points was laid onto each individual image taken and the intersections with the points and the particle boundaries were counted (see Figure 3.2).

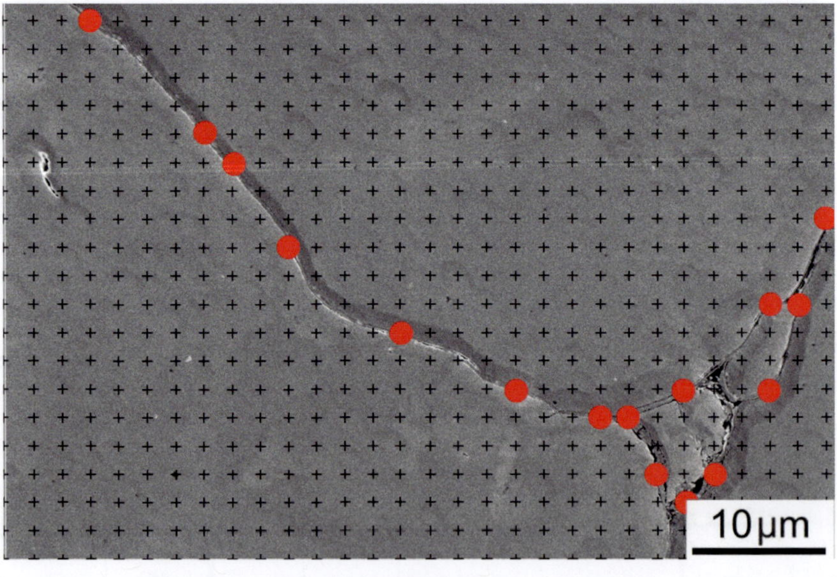

Figure 3.2: For the point count method, a grid of 30 x 20 points was laid on an SEM image of the SMCs. The red points are the intersections with the particle boundaries. Then, the type of boundary was classified.

3.6 Electron Backscatter Diffraction (EBSD)

Electron Backscatter Diffraction (EBSD) is a powerful characterisation tool to analyse grain orientations, internal textures in grains or phase identification. In case of the SMCs, the internal structure of the iron particles was analysed and EBSD was applied during the preparation procedure for microtensile test specimens (see section 3.11). An FIB/SEM dual beam microscope (JIB-4601F, JEOL, Japan) equipped with an EBSD system (TSL DVC5) was used. The sample is tilted to 70° from horizontal or mounted on a pre-tilted EBSD holder. The scattered electrons hit a phosphorous screen resulting in a specific fluorescent pattern. A camera and optics system then detects the pattern formed on the screen. The orientation was determined using automated beam scanning, with step sizes of 0.2 and 4 μm at an acceleration voltage of 20 kV. The EBSD analysis was carried out using the TexSEM Laboratories orientation imaging microscopy (OIM) software (v7.1.0). A clean-up procedure was applied to all EBSD images to adjust single points with misorientations of more than 5° to their neighbours. [41]

3.7 Electron Microscopy and Focused Ion Beam Milling

Different DualBeam® microscopes combining scanning electron microscopy (SEM) and focussed ion beam (FIB) were used for imaging and cutting cross sections: a Nova NanoLab 200, a Helios NanoLab 650 and a Scios (FEI Company, Hillsboro, Oregon, USA). Some of the ion beam polished images were taken with a Zeiss Supra 55 (Zeiss AG, Oberkochen, Germany). [41] TEM lamellas are prepared by FIB milling in the Nanolab 200 and 650 according to the Lift-Out Method [103]. This method requires a micromanipulator to lift out the FIB-milled TEM lamella: in the Nanolab 200 it was a Kleindiek MM3 and in the Nanolab 650 an OmniProbe 100. *The microcompression and microcantilever samples were prepared using a focused ion beam (FIB) platform (Helios Nanolab 650 Dual Beam, FEI Company, Hillsboro, OR, USA). Pillars and microcantilevers with and without particle boundaries were machined in multi-*

step processes. The pillars had square cross sections of 5 μm x 5 μm and a height of about 20 μm. Prior to FIB-machining, a thin protective Pt layer was deposited using electron-beam-assisted deposition. Then, trenches were cut at 65 nA around the Pt square defining the size of the pillar cross section. Next, the actual pillar shape was cut at 9 nA beam current. The final milling step was performed at 0.7 nA. The taper angle was in the range of 0.6 - 2.4 °. The aspect ratio – pillar height divided by width – was 2.8 - 4.8 to minimize triaxial stress states and avoid buckling [104, 105]. Microcantilever beams with triangular cross sections were prepared similar to the process described in [68]. Prior to the coarse milling, a protective carbon layer was deposited by electron-beam assisted deposition at a beam current of 65 nA. For the fine milling, a beam current of 9 nA was used, while the final polishing of the cantilever beam side wall was performed at 0.7 nA. The final cantilevers had a length of 30 μm and a triangular cross section with a width of 2.8 - 5.1 μm. [106]

3.8 Transmission Electron Microscopy and Energy-Dispersive X-Ray Spectroscopy

*A Transmission Electron Microscope (TEM) (**Tecnai G² F20 X-TWIN,** FEI Company, Hillsboro, Oregon, USA) operated at 200 kV was used to take images in the scanning TEM (STEM) mode. [41] The samples were analysed by SEM and TEM before and after deformation. Energy Dispersive X-Ray Spectroscopy (EDX, EDAX S-UTW) was performed to identify individual layers of the complex particle boundaries of the SMCs.* [106]

3.9 Nanoindentation

The indentation tests were conducted in a G200 XP nanoindenter (Keysight Technologies, Inc., Santa Rosa, CA, USA) with a diamond Berkovich tip using the Continuous Stiffness Measurement (CSM) technique. [41] In the CSM method, a small sinusoidal load signal is superimposed onto the normal load signal. This allows the measurement of contact stiffness at any point along the loading curve and not just at the point of unloading as in the conventional measuremen [43, 49]. The tip was calibrated following the standard method [38]. Three sets of experiments were performed.

In set 1, the influence of the polishing conditions on the nanoindentation measurements was examined on two different specimens. The first one was polished mechanically up to step 7a (Table 1.1), while the other one was ion beam polished (following procedure D, Table 3.1). The samples were then indented to 900 nm depth at a constant indentation strain rate of 0.05 s^{-1}. 100 indents were made inside the particles of the SMC. Young's modulus and hardness were determined for the bulk iron particles. [41]

In set 2, indents were placed along a selected particle boundary of an ion beam polished sample (procedure B, Table 3.1). The indentation depth was 300 nm resulting in a lateral size on the surface of approximately 2.3 µm, which was greater than the maximum thickness of the particle boundaries. [41]

In set 3, indents were placed along particle boundaries over a full specimen cross section of a fully oxidized sample in order to identify the individual hardness of each type of boundary. The indentation depth was 300 nm.

3.10 Microcompression and Microcantilever Bending

The compression and bending tests were performed using a nanoindenter (G200, Keysight Technologies, Santa Rosa, California). The microcompression pillars were deformed with a flat punch diamond tip of 10 µm diameter. All experiments were carried out with a constant displacement rate of 10 nm/s to a predefined displacement. [106]

As described in section 1.3.2, there are several aspects that have to be considered because microcompression testing is not standardised. The influence of the taper and the substrate are taken into account when calculating the stress and strain of the microcompression pillars. The pillar strain cannot directly be measured in a microcompression experiment [62, 107, 108]. The total displacement can be described as

$$\Delta x_{total} = \Delta x_{ind} + \Delta x_{pillar} + \Delta x_{sub} \tag{3.1}$$

where Δx_{total} is the indentation depth and Δx_{ind}, Δx_{pillar} and Δx_{sub} are the deformation in the indenter, pillar and substrate respectively. The pillar deformation is given by

$$\Delta x_{pillar} = \Delta x_{total} - \frac{\left(1-\upsilon_{sub}^2\right)\cdot P}{2\cdot\beta\cdot E_{sub}\cdot r_{base}} - \frac{\left(1-\upsilon_{ind}^2\right)\cdot P}{2\cdot\beta\cdot E_{ind}\cdot r_{top}} \tag{3.2}$$

with

$$P = \pi \cdot r_{ind}^2 \cdot p \tag{3.3}$$

where υ_{sub} and υ_{ind} are the Poisson's ratio for the substrate and the indenter, P is the applied force on top of the indenter and p is the pressure, β is a correction factor, E_{sub} and E_{ind} are the Young's moduli of the substrate and the indenter and r_{base} and r_{top} are radii of the top and base surface of the pillar. The correction factor β is determined by finite element calculations of the load-displacement characteristics of elastic half spaces deformed by flat ended punches with circular (β = 1.000), triangular (β = 1.034), and square cross sections (β = 1.012) [47]. An indenter tip is typically made of diamond with a Young's modulus of 1141 GPa and a Poisson's ratio of 0.07. The strain ε can be calculated as follows:

$$\varepsilon = \left(1 - \frac{\Delta x_{pillar} - \Delta x_{total}}{\Delta x_{pillar}}\right) \cdot 100\% \tag{3.4}$$

The stress σ is determined by:

$$\sigma = \frac{P}{A} \tag{3.5}$$

where A is the area of the top surface of the pillar [62].

The microcantilever beams were deformed in the nanoindenter with a diamond wedge tip with a length of 8 µm and an opening angle of 60° at a constant displacement rate of 10 nm/s to a predefined displacement.

Simple beam theory known from the macroscopic bending test can be also applied to the microscale if some requirements are fulfilled [68, 109, 110]. The aspect ratio (loaded length / microcantilever width) of the microcantilevers should be greater than 6 because otherwise the linear beam theory is not valid and the Young's modulus is severely underestimated [68]. The evaluation of Young's modulus E is adapted from [68]. It can be calculated by

$$E = \left(\frac{P}{d}\right) \cdot \frac{L^3}{3 \cdot I_y} \tag{3.6}$$

with

$$I_y = \frac{bh^3}{36} \tag{3.7}$$

where d is the deflection of the cantilever at the loading point, L is the distance between the loading point and the fixed end, h is the cantilever thickness and b is the cantilever width. In order to determine Young's modulus, the sizes of the microcantilevers and the value $\left(\frac{P}{d}\right)$ are needed for equation (8) and (9). $\left(\frac{P}{d}\right)$ is determined by a least square fit in the elastic region of the loading and unloading curve.

3.11 Microtensile Testing

The microtensile test specimens were prepared and tested similar to the process and method described in [74, 81, 82]. They were cut from fully oxidised SMCs into approximately 250 µm thick slices using a diamond wire saw without cooling. Both surfaces were then thinned by grinding with SiC paper down to P2000. The final thinning step to a thickness of 25 µm was performed using a cross section ion beam polisher (SM-09010, JEOL, Japan) resulting in a polished surface shaped in a half-circle. Specimens with a gauge section of ~20 µm x 25 µm x 50 µm were fabricated using focused ion beam milling (JIB-4601F, JEOL, Japan). EBSD measurements had to be performed in order to determine the crystal orientation of the gauge sections of the specimens. The test was performed with a micro-gluing grip [70] at a crosshead speed of 0.1 µm/s which was equivalent to a strain rate of 0.002 s^{-1}. The fracture surfaces were examined by SEM.

3.12 Macroscopic Four-Point Bending

A single-edge bend sub-size specimen KLST (Kleinstprobe) with nominal dimensions of 3 mm x 4 mm x 27 mm, a notch depth of 1 mm, a notch angle of 60° and a root notch radius of 0.1 mm was cut by electrical discharge machining (EDM) [111] (see Figure 3.3). The dimensions were chosen due to the limited amount of material. From each fully oxidised sample, eight specimens per bar were cut. From each partially oxidised sample, also eight specimens were cut. The notch was placed either at the surface of the sample (specimen A) or in the core region of the sample (specimen B) (see Figure 3.3). The surface of selected specimens was ground and polished down to 3 µm with diamond suspension for observing the crack [41], while the others were tested without further surface preparation. All of the specimens were then further notched by the razor blade method [112] using no diamond paste but with oil as a lubricant. The crack length and radii were measured optically with a light microscope (Eclipse LV150, Nikon, Tokyo, Japan). The radii obtained with this method were 10-23 µm and the final notch depth was 1.1-1.2 mm.

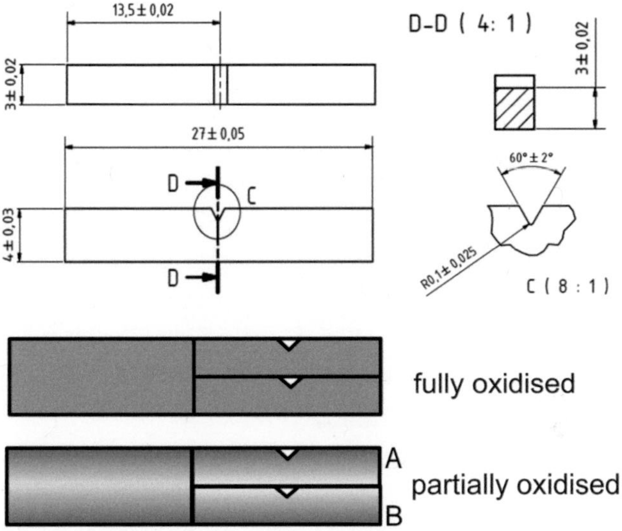

Figure 3.3: Dimensions of a KLST specimen and specimen placement in a fully oxidised and partially oxidised sample.

The notched specimens were tested in a four-point bending clamping fixture with inner and outer spans of 6 mm and 12 mm, respectively. The specimens were tested under displacement control in a Universal Testing System (Type 10, 50 Hz, UTS Test Systeme, GmbH, Ulm, Germany) at a cross-head speed of 1.5 mm/min. In order to observe the crack propagation, a force-controlled Instron 1362 (Instron, Canton, MA, USA) with 10 N/s and 80 N/s corresponding to 0.04 mm/min and 2.00 mm/min, respectively, was used. Videos of the notched area were recorded for selected specimens with a CCD video camera in order to correlate the propagation of the crack with the load-displacement curve.

The fracture toughness was determined using the single etch v-notched beam (SEVNB) method [88] as suggested in [113]. A sharp notch is introduced by polishing the notch root with a razor blade and impregnated with diamond paste. In order to determine the crack length *a*, the fracture surface has to be measured at three different points and a mean value is calculated (see Figure 3.4).

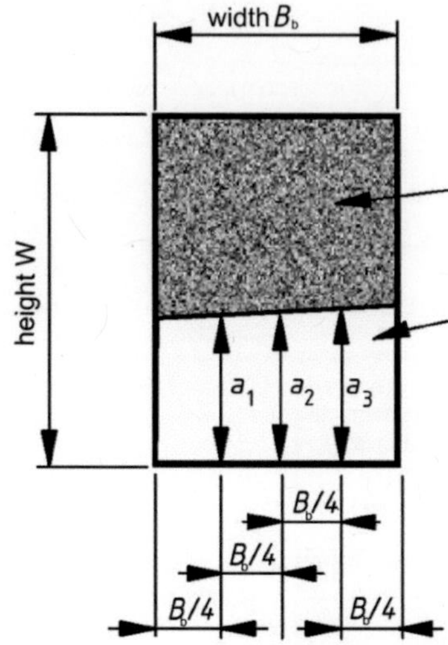

Figure 3.4: Schematic of a fractured specimen with dimensions and positions of a_1, a_2, and a_3 for determining the crack length (adapted from [89]).

The mean relative crack length α in equation (12) must be between 0.2 and 0.3 and fulfil equation (11) [89].

$$a = \frac{(a_1 + a_2 + a_3)}{3} \tag{3.8}$$

$$\frac{(a_{\max} - a_{\min})}{a} \leq 0.1 \tag{3.9}$$

$$\alpha = \frac{a}{W} \tag{3.10}$$

where a is the crack length, a_{max} is the largest value of a_1, a_2 and a_3, a_{min} is the smallest value of a_1, a_2 and a_3, α is the relative crack length and W is the height of the specimen

In metals, only a preliminary stress intensity factor K_Q can be determined from the experiments. K_Q is only equal to the critical stress intensity factor K_{IC} if

$$\begin{Bmatrix} a \\ W-a \\ B_b \end{Bmatrix} \geq 2.5 \cdot \left(\frac{K_Q}{\sigma_y} \right)^2 \tag{3.11}$$

where B_b is the width of the specimen, K_Q is the preliminary stress intensity factor and σ_y is the yield strength [114]. Unlike in metallic materials, it is possible to directly determine a K_{IC} for ceramic materials if the critical flaw size is significantly smaller than the grain size of the tested material [90].

For a four point bending test, the critical stress intensity factor K_{IC} is [89]:

$$K_{1C} = f\left(\frac{a}{W}\right) \cdot \left[\frac{P_{max} \cdot (S_0 - S_1) \cdot 10^{-6}}{B \cdot W^{3/2}} \right] \cdot \left[\frac{3 \cdot \alpha^{1/2}}{2 \cdot (1-\alpha)^{3/2}} \right] \tag{3.12}$$

with a geometry factor [89]:

$$f\left(\frac{a}{W}\right) = 1.9887 - 1.32 \cdot \alpha - \frac{(3.49 - 0.68 \cdot \alpha + 1.35\alpha^2) \cdot \alpha \cdot (1-\alpha)}{(1+\alpha)^2} \tag{3.13}$$

4 Sample Preparation

4.1 Grinding and Polishing

Figure 4.1 shows heavy corrosion stains between polishing steps. During grinding, no corrosion stains were observed. During polishing, however, they regularly showed up at different stages of corrosion. The stains could usually be removed by iterating the cleaning procedure. If not, the polishing step had to be repeated.

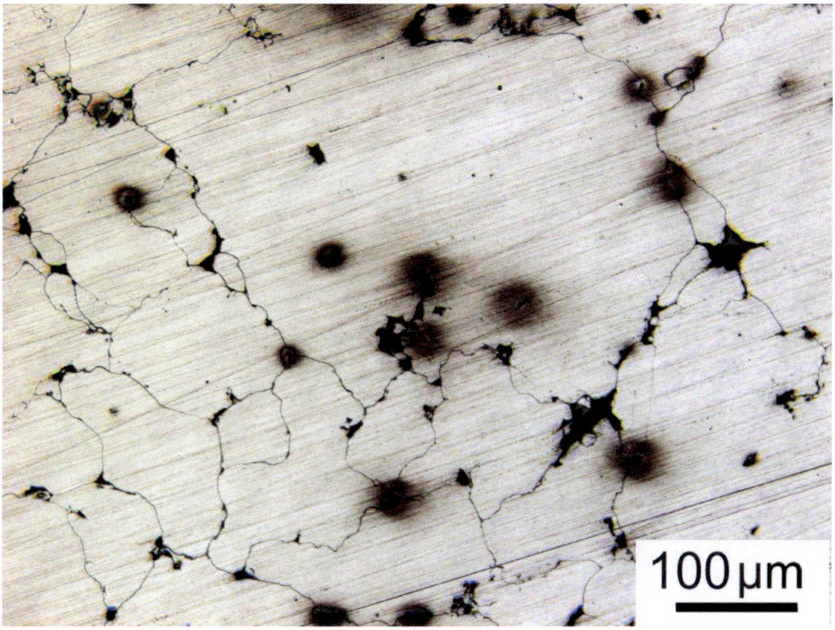

Figure 4.1: Corrosion stains after polishing. Sometimes the stains could be removed by cleaning several times. Sometimes the polishing step had to be repeated.

Mechanical polishing only (up to step 7 in Table 3.3) is not sufficient to reveal the real microstructure of the particle boundaries. The specimen shown in Figure 4.2a was polished up to step 7a. Scratches as well as imprints of the diamond grains of the polishing suspension are clearly visible. The surface quality clearly improved when step 7b instead of step 7a was the final polishing step (Figure 4.2b). In addition, slight etching of the iron particles can be seen in Figure 4.2b. The particle boundary, however, appears unaffected compared to the previous polishing procedure (step 7a). Both polishing steps reveal the individual layers of the boundary. One layer was more severely attacked by the polishing process than the others resulting in a relief of the particle boundary. Apparently, the larger macro-pores had been filled by the infiltration process, whereas the very small pores along the boundary layers remained unfilled. After mechanical polishing, these small pores appear smeared up to some extent. [41]

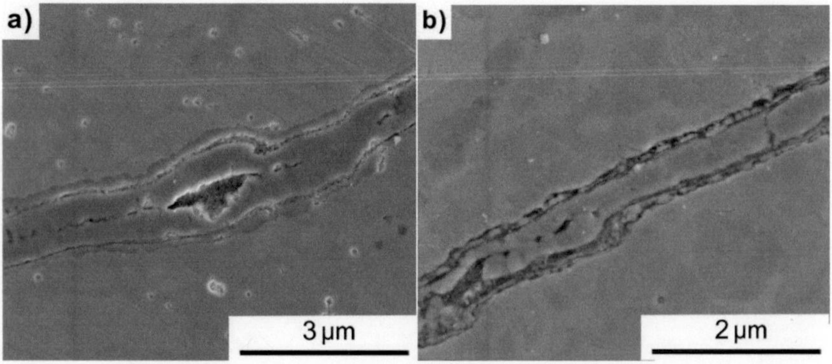

Figure 4.2: Scanning electron micrographs of the SMC samples after different polishing procedures: a) Mechanical polishing (step 7a): Scratches and imprints in the iron particles of the polishing abrasive can be seen. The boundary exhibits a surface relief. b) Mechanical polishing (step 7b): The particles show fewer scratches and imprints. The boundary still exhibits a surface relief. [41]

4.2 Ion Beam Polishing

4.2.1 Planar Ion Beam Polishing

Figure 4.3a shows a specimen polished following procedure A (Table 3.4). Even with a cleaning step, ion beam polishing did not remove all scratches left after the mechanical preparation. The boundaries, however, did not show a surface relief anymore and the small pores were also laid open. Procedures B and C gave even better results concerning the surface quality of the iron particles. Procedures C and D resulted in directional over-etching, locally along the particle boundaries (see Figure 4.3b), due to the lateral movement of the specimen during ion beam polishing. Although the over-etching caused a more irregular surface, the ion beam polishing in C and D gave the best and most reproducible overall results. In Figure 4.3b (procedure D), the orientation contrast within the iron particles is clearly visible. This orientation contrast disappeared with increasing surface corrosion. [41] The voltage and polishing time were increased in Procedure E resulting in a strongly over-etched specimen not only along the particle boundaries but also at the particle surfaces (Figure 4.3c).

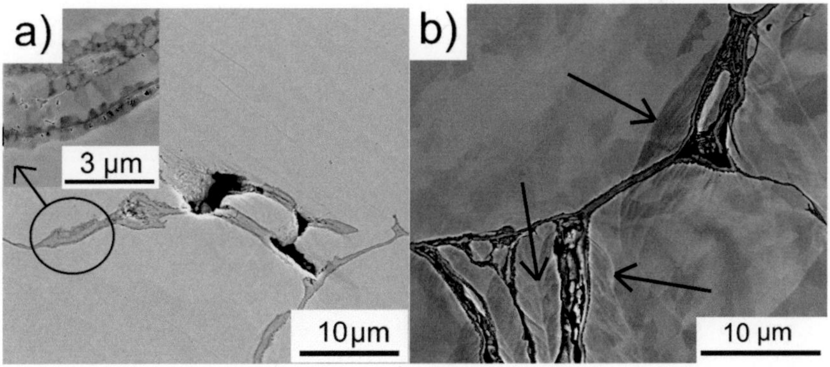

(Figure 4.3 continued on next page)

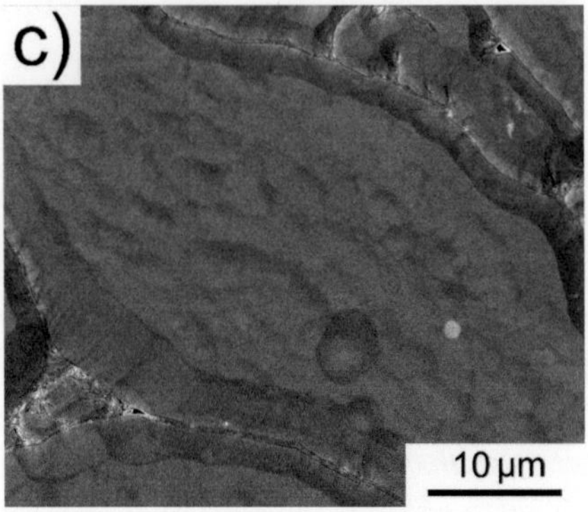

Figure 4.3: Scanning electron micrographs of the SMC samples after different polishing procedures: a) Ion beam polishing (procedure C): The surface relief of the boundary was removed. The individual layers of the boundary as well as the pore structure are clearly visible (see enlargement). b) Ion beam polishing (procedure D): Directional over-etching occurred at the particle boundaries (marked by arrows), while the particle and the boundaries (see smaller image) were otherwise smoothly polished. c) Ion beam polishing (procedure D): strong over-etching of another specimen. (Figure 4.3 a and b from[41])

(SEM micrographs in Figure 4.3a and b courtesy of Dr. Marcus Müller)

4.2.2 Cross-Sectional Ion Beam Polishing

The specimen surface preparation is critical for EBSD measurements since the real microstructure may not be exposed. The SEM micrograph in Figure 4.4a shows the ion beam polished portion of the cross section shaped in a half-circle after mechanical polishing. As shown in the coarse EBSD scan of the whole cross-sectional area with a step size of 4 µm (Figure 4.4b), procedure CS-1 (Table 3.4) leads to over-etching particularly at the edge of the specimen. Large areas at the edge could not be indexed, and, hence, the identification of the crystal orientation failed (black regions in the Inverse Pole Figure (IPF) map in Figure 4.4b). *In Figure 4.4c, however, the indexing of a properly polished sample (CS-2 in Table 3.4) was more successful with unidentified orientations mainly at the particle boundaries (see Figure 4.4d).* [41]

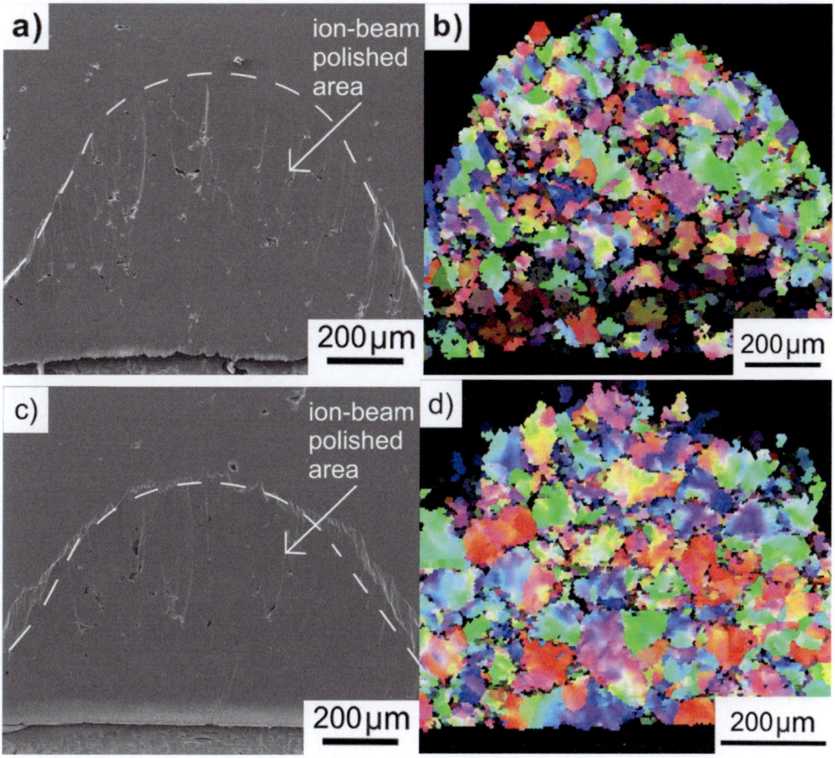

Figure 4.4: EBSD images showing the crystal orientation of the iron particles. a) The specimens were prepared by ion beam-polishing (SEM), b) The indexing success rate varied with the polishing procedure. Over-etching (CS-1) led to unindexed areas along the sample edge as can be seen in the IPF map with a step size of 4 µm. c) The cross-sectional ion beam polished surface had different polishing conditions (CS-2) d) Procedure CS-2 results in smaller unindexed areas. (Figure 4.4a, b and d from [41])

(SEM and EBSD micrographs courtesy of Dr. Yoji Mine, Kumamoto University)

4.3 Deformation Layer

FIB-milled cross sections of the differently polished samples were prepared without tilting the sample to an angle of 52°. The specimen shown in Figure 4.5a was polished mechanically (step 7a in Table 3.3). The thickness of the deformation layer was approximately 311 nm as indicated by the contrast changes of the sub-cell structure in the particles. As shown in Figure 4.5b and Figure 4.5*c, the thickness of the deformation layer was reduced to 163 nm after ion beam polishing (procedure B in Table 3.4) or even completely removed (procedure D), respectively.* [41]

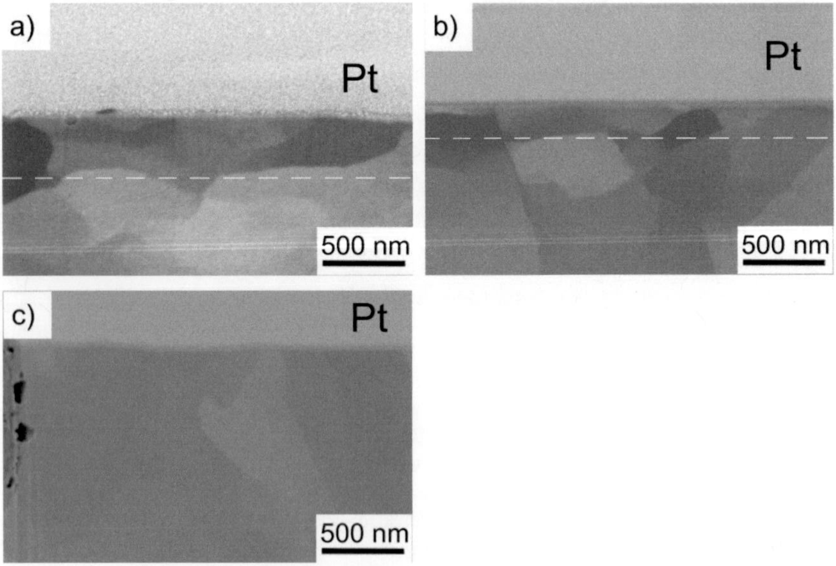

Figure 4.5: SEM micrographs of FIB cross sections prepared at an angle of 52° to the surface normal [41] a) a mechanically polished specimen (step 7a) showing the deformation layer of ~245 nm thickness with clearly visible subgrains, b) an ion beam polished specimen (procedure B) with a deformation layer of ~129 nm thickness, and c) of an ion beam polished specimen (procedure D) without deformation layer. The cross sections were imaged at 2 mm working distance with a backscatter detector (In-lens Trinity 1, FEI Company, Hillsboro, Oregon, USA). [41]

(SEM micrographs courtesy of Thomas Kreuter)Material Characterisation

5 Material Characterisation

5.1 Powder Characterisation of Somaloy 500 and 700 3P

Somaloy 500 and Somaloy 700 3P powders in their original state were analysed in the SEM. The differences between those two powders were the additional binder layer of Somaloy 700 3P and a different particle size distribution.

Figure 5.1a shows an overview of the examined Somaloy 500 powder with different sizes of particles. The particle in Figure 5.1b had the typical water-atomised random surface with some unidentified white inclusions. A cross section of this particular particle was prepared using FIB-cutting. The channelling contrast in Figure 5.1c indicates a grain boundary within the particle. There was also a small pore within the particle. The phosphorous containing coating could not be resolved in the SEM (see Figure 5.1d).

Figure 5.2a shows the particles of the examined Somaloy 700 3P powder. There is an obvious size difference of the particles compared to the Somaloy 500 and they also include some white inclusions of unknown origin (Figure 5.2b). On the cross section of this particular particle (Figure 5.2c), it seemed to be single crystalline and did not include any pores. In Figure 5.2d, the binder layer on the surface of the particle is shown but the phosphorous layer was again not resolvable by SEM.

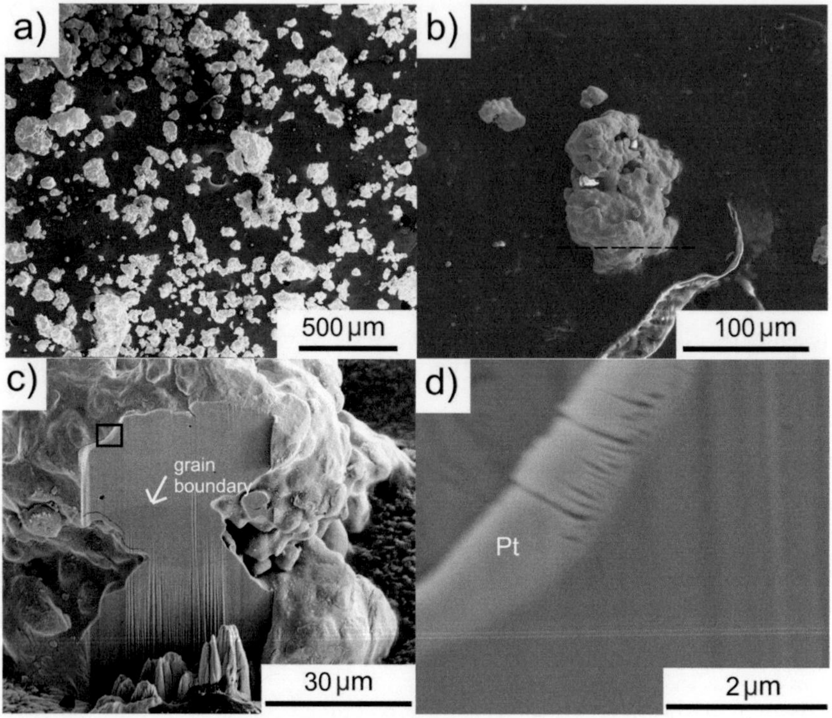

Figure 5.1: Somaloy 500 a) An overview of the powder, b) The powder particle was cut with FIB at the marked position c) The cross section through the particle shows a grain boundary, d) The surface of the particle (marked in c)) with a platinum (Pt) coating.

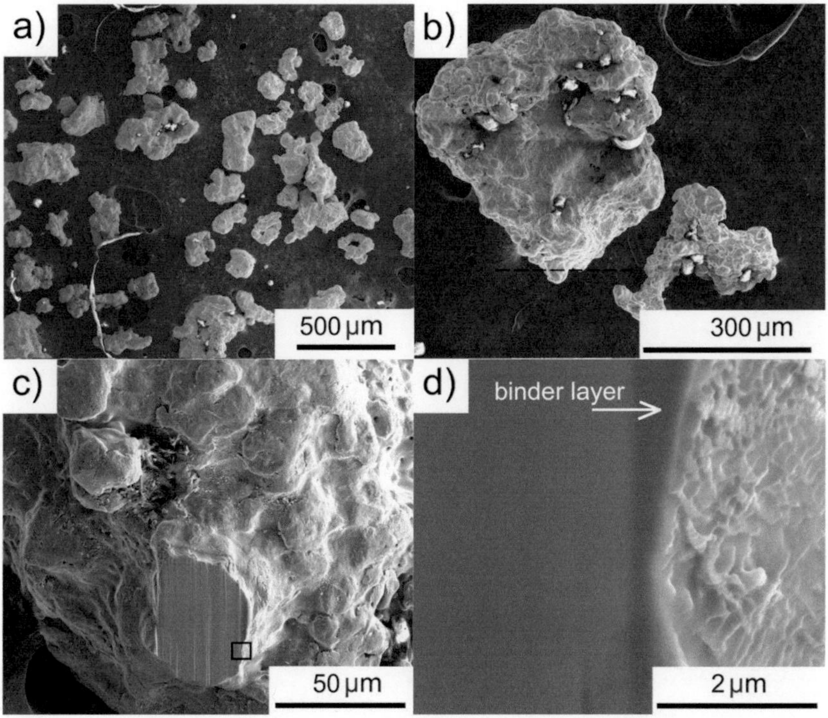

Figure 5.2: Somaloy 700 3P a) An overview of the powder, b) The powder particle was cut with FIB at the marked position, c) The cross section through the particle, d) The surface of the particle (marked in c)) showing the binder layer.

5.2 EBSD Measurements

The particles were characterized by ESBD scans with a step size of 0.2 µm. Figure 5.3 shows inverse pole figure (IPF) maps combined with image quality (IQ) maps and each image shows two particles and a particle boundary. Figure 5.3a shows that the structure of the particle boundary could not be resolved. Within the two particles, there were only minor colour orientation changes indicating the subgrain structure of the particles seeming to be single-crystalline. In Figure 5.3b however, there are clearly two grains in the particles visible in the scanned area. Each grain again shows a subgrain structure. In Figure 5.3c, the right particle even contains 3 individual grains in the scanned area and the left particle contains two grains. In Figure 5.3d, there is a pronounced colour variation indicating subgrains and a grain boundary in the left particle.

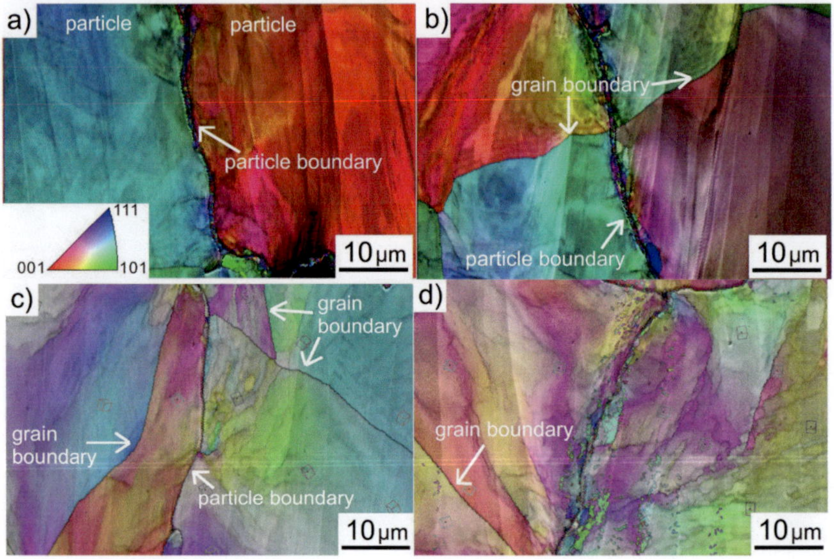

Figure 5.3: Higher magnification EBSD scans (combined IPF and IQ maps) with a step size of 0.2 µm: a) the particle boundary could not be resolved by ESBD scans but subgrains are detectable. The particles seem to be single-crystalline, b) there is a grain boundary in both particles and a subgrain structure in each grain, c) there are two grain boundaries in the left side particle and three grain boundaries in the right particle, d) the particles show a pronounced subgrain structure and the left particle contains a grain boundary. (Figure 5.3a from [41])

(EBSD micrographs courtesy of Dr. Yoji Mine, Kumamoto University)

5.3 Microstructural Analysis in the TEM

Figure 5.4 shows a STEM image of a boundary and an iron particle. Within the particle, a cellular subgrain structure with high dislocation density can be observed.

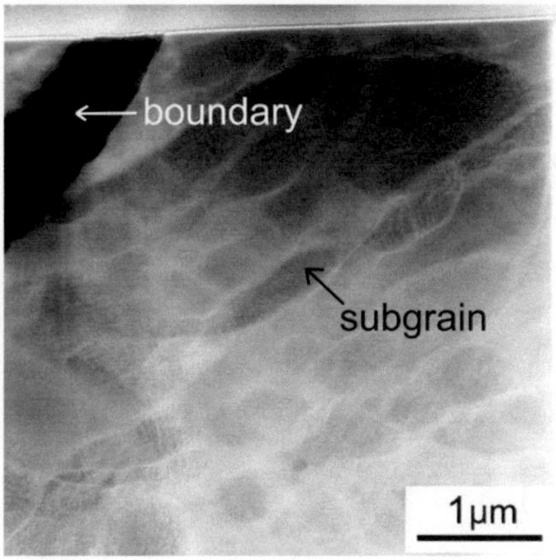

Figure 5.4: STEM image of an iron particle with a boundary. Cell structures of subgrains are clearly visible. [41]

As shown in Figure 5.5, the particle boundaries of the SMC are very complex. Depending on the space between the iron particles during oxidation, the boundaries grew different layers of oxidized iron and phosphorous. Four general types were observed. Figure 5.5a shows the initial state of the boundary without oxidation and only a phosphorous layer of ~30 nm thickness. Figure 5.5b shows a thicker boundary (40-60 nm) with pores and traces of iron oxide in the middle (type-2 boundary). Type-3 particle boundaries exhibit an iron oxide layer with a thickness of several tens of nanometres (in this case even ~350 nm) in addition to a phosphorous layer, as shown in Figure 5.5c. Interface pores between the particle and the phosphorous layer (~62 nm thick) as well as larger pores in the middle of the iron oxide layer can be seen. In case of the type-4 boundaries (Figure 5.5d), the first two layers adjacent to the particle are iron oxide and phosphorous, respectively. [106] The iron oxide layer also contains small pores.

As the SMC is only pressed and not sintered, the structure is generally porous. During compression, the water-atomised particles are jammed resulting in different sizes of pores. The pores offer a lot more space for boundary growth than the later boundary layers. The growth of boundary layers does not depend on the pore sizes but on the permeability for the water vapour during annealing.

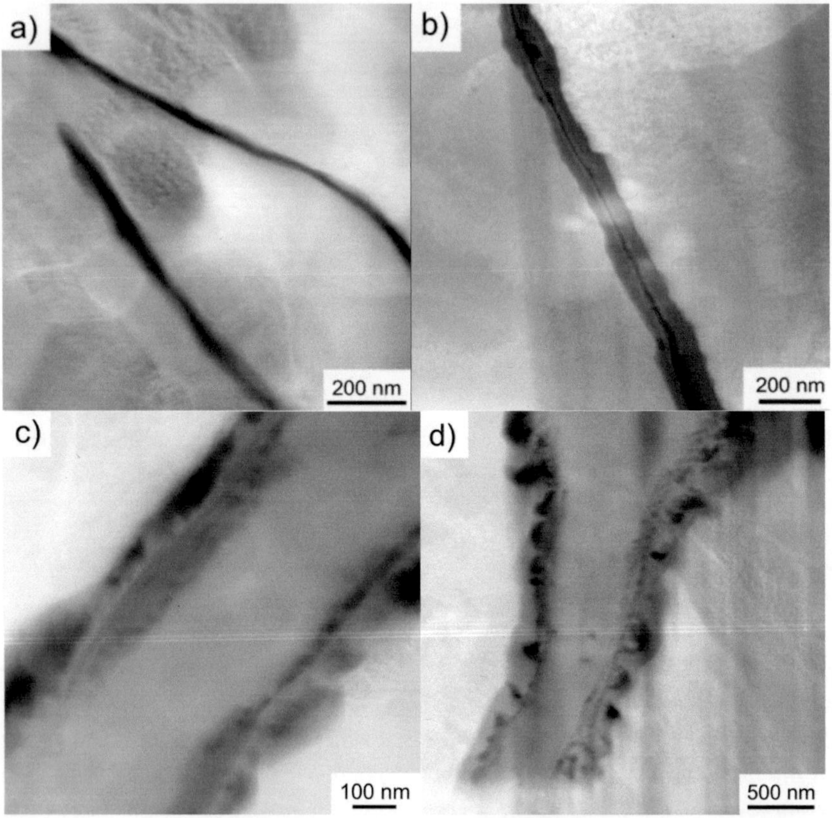

Figure 5.5: STEM micrographs of a) a type-1 boundary with a phosphorous layer of about 20 nm thickness, b) a type-2 boundary with a thicker phosphorous layer and traces of iron oxide, c) a type-3 boundary containing a thick iron oxide layer besides the phosphorous layer, and d) a type-4 boundary with a porous iron oxide layer, a thin phosphorous layer and a middle iron oxide layer including a pore. (Figure 5.5a, b and c from [106])

This means that even small pores are not necessarily filled up if the water vapour cannot reach them (Figure 5.6a and b). The growth of boundary layers at the pore surface is in general the same as described before (Figure 5.6c). However, they grow randomly within the pore, filling it up and making it very complex (Figure 5.6d).

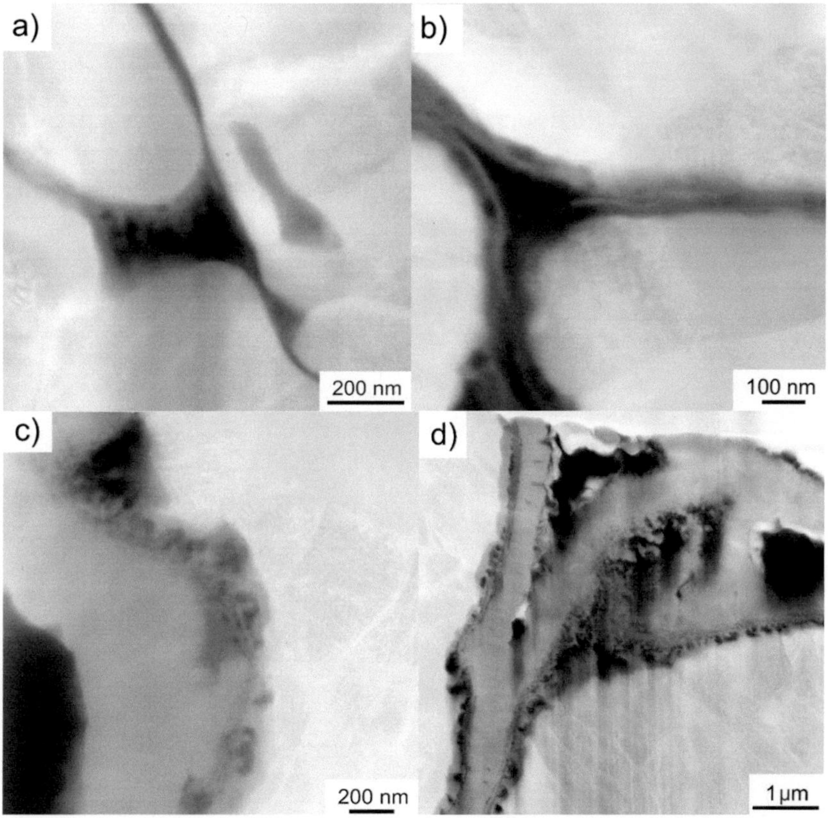

Figure 5.6: STEM micrographs of different types of pores a) a pore surrounded by a type-1 boundary b) a pore enclosed by a type-2 boundary with a thicker phosphorous layer and the beginning growth of an iron oxide layer c) a large pore enclosed by a type-3 boundary: with an iron oxide layer in the middle surrounded by a phosphorous layer. d) a complex filled pore of a type-4 boundary with an iron oxide layer at the particle side and a phosphorous layer followed by an iron oxide layer which also filled parts of the pore. *(*Figure 5.6*d from [106])*

Figure 5.7 shows EDX line scans for several types of particle boundaries. The line scan in Figure 5.7a is across a type-1 boundary consisting of oxygen, iron and phosphorous. The amount of iron decreased and the amount of oxygen and phosphorous increased compared to the pure iron particle. The thickness of the boundary is about 20 nm.

Figure 5.7b shows the transition of a type-2 into a type-3 boundary. The line scan across this boundary shows both the phosphorous layers as well as the thin iron oxide layer in the middle. The phosphorous layers are 50 nm (layer 1) and 70 nm (layer 3) thick and the iron oxide layer is 80 nm (layer 2) thick. The amount of oxygen is nearly constant over the whole boundary, only the amount of iron decreases in the phosphorous layer.

As shown in Figure 5.7c, there is a continuous decrease of iron content when following the line scan starting in the iron particle and increasing oxygen content over a distance of ~ 131 nm (layer 1 in Figure 5.7c). The second layer is the phosphorous layer containing the least amount of iron and highest amount of oxygen (layer 2, ~ 72-156 nm). The layer in the middle of the boundary (layer 3, ~ 366 nm) consists of iron oxide with a higher amount of iron than oxygen [41].

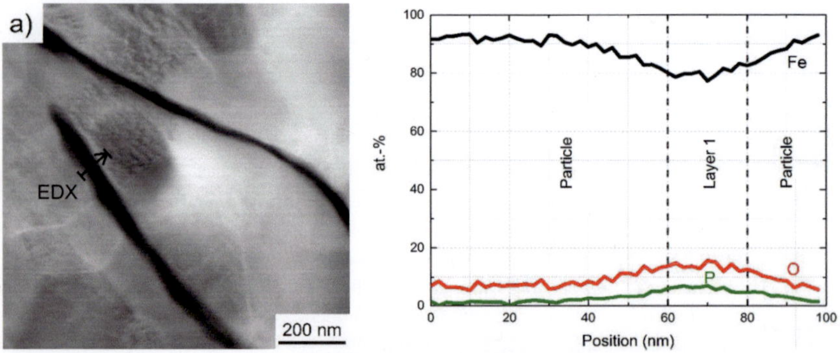

(Figure 5.7 continued on next page)

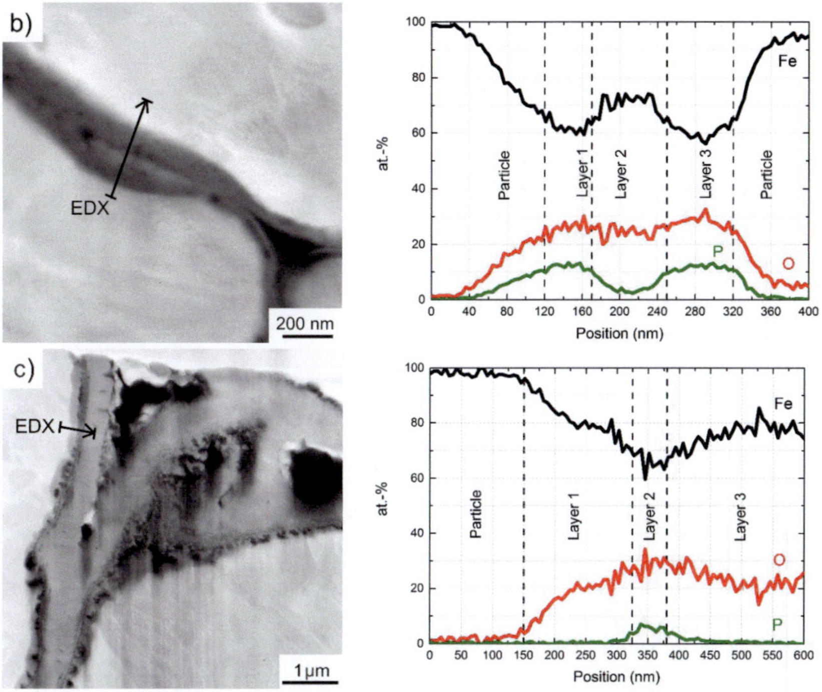

Figure 5.7: STEM micrographs and their corresponding EDX line-scans for a) a type-1 boundary consisting of a phosphorous layer, b) a type-2 boundary in transition to a type 3 boundary and c) a type-4 boundary with two layers of iron oxide and phosphorous, and a middle layer consisting of iron oxide. *(Figure 5.7d from [106])*

5.4 Quantitative Microstructural Analysis

Figure 5.8 shows the percentage distribution of all types of particle boundaries of a cross section of a partially oxidised sample. The distribution of type-1 and type-2 boundaries is similar with a higher percentage of these particle boundaries in the core region (Figure 5.8a and b). There are generally more type-2 boundaries than type-1 boundaries. The number of type-3 boundaries is higher outside the core than inside of the core (Figure 5.8c). Almost no type-4 boundaries are in the core but there are more of type-4 boundaries outside the core (Figure 5.8d).

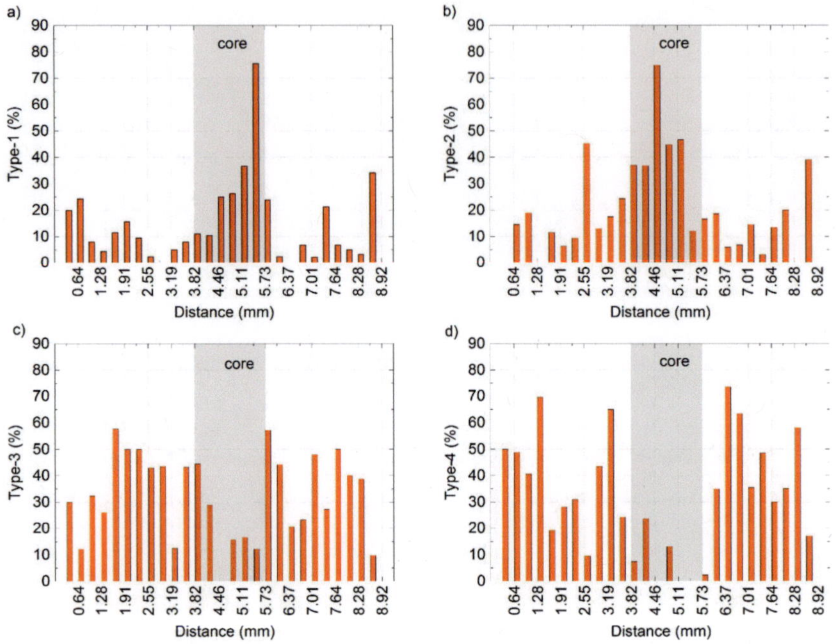

Figure 5.8: Boundary type distribution along a cross section of a partially oxidised sample: a) There is a high number of type-1 boundaries in the core, b) There are generally more type-2 than type-1 boundaries and there is also a higher number of type-2 in the core, c) the number of type-3 boundaries is generally higher outside the core region and there are less boundaries in the core, d) There are very few type-4 boundaries in the core region.

Figure 5.9 shows the boundary type distribution for a full sample cross section of a fully oxidised sample. There are generally very few type-1 boundaries in the whole sample (Figure 5.9a). The amount of type-2 is higher than the number of type-1 boundaries with a locally higher concentration on the left side of the sample (Figure 5.9b). There are even more type-3 boundaries with a slightly increasing number of particle boundaries to the right side of the sample (Figure 5.9c). The type-4 boundary shows the highest distribution with a high number of boundaries at the core region and a somewhat smaller number towards the edges of the sample (Figure 5.9d).

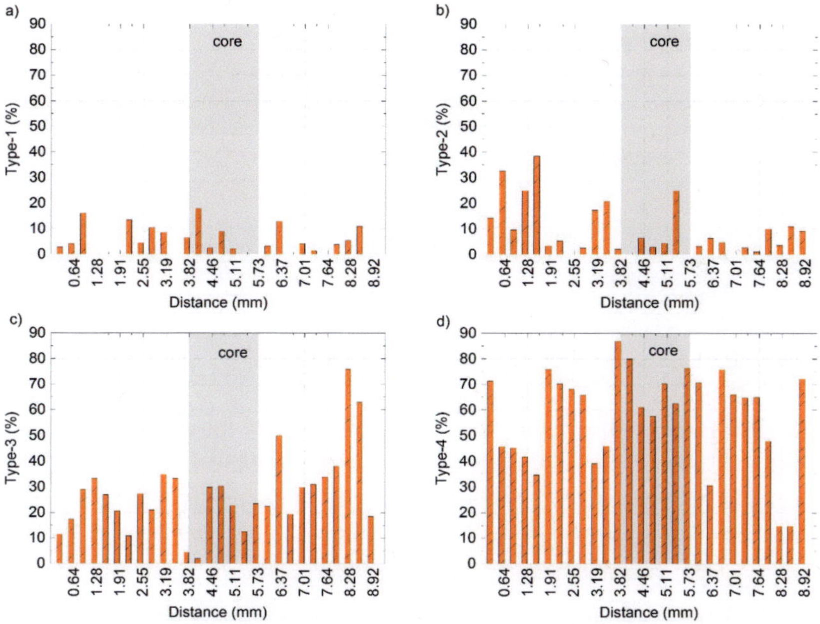

Figure 5.9: Boundary type distribution along a cross section of a fully oxidised sample: a) There are very few type-1 boundaries in the whole sample, b) There are generally more type-2 than type-1 boundaries with a locally higher number at the left side of the sample, c) the number of type-3 boundaries is generally higher and slightly increasing to the right side of the sample, d) The type-4 boundary shows a high number of boundaries at the core region and a little lower number towards the edges of the sample.

Table 5.1 shows the results of the Point Count Method. In total, the percentage of particle boundaries within the analysed surface was 1.5% and 1.2% for fully and partially oxidised SMCs respectively. Fully oxidised specimens contained 10.3% of type-1 and type-2 boundaries, 27.8% of type-3 boundaries and 59.5% of type-4 boundaries. Considering partially oxidized SMCs, the number of type-1 and type-2 boundaries increased by 10% and 14%, respectively. The number of type-3 boundaries remained almost the same and the number of type-4 boundaries dropped to 32.6%.

Table 5.1: Distribution of all types of particle boundaries in the SMCs for fully and partially oxidised specimen.

	Fully oxidised SMC	Partially oxidised SMC
boundaries in SMC	1.5 %	1.2 %
type-1 boundary	4.9 %	15.0 %
type-2 boundary	5.4 %	19.5 %
type-3 boundary	27.8 %	33.0 %
type-4 boundary	59.5 %	32.6 %

The SEM micrographs in Figure 5.10 show the core of fully and partially oxidised SMCs respectively. As already shown in the Point Count Method, the core of fully oxidised SMC contains only type-4 boundaries. This can be also seen in the SEM micrograph in Figure 5.10a. Similarly, the core of partially oxidised SMC contains many type-1 boundaries as shown in Figure 5.10b.

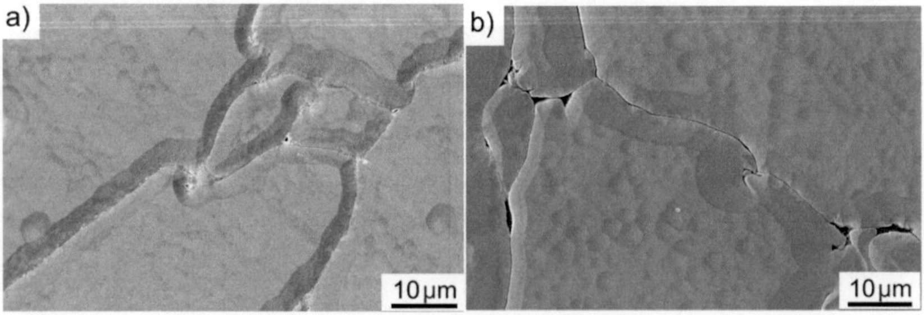

Figure 5.10: a) SEM micrograph taken at the core of a fully oxidized specimen. There are mainly type-4 boundaries. b) SEM micrographs taken at the core of a partially oxidized specimen. There are only type-1 boundaries in this image.

6 Micromechanical Testing

6.1 Nanoindentation of Pure Iron Particles

Figure 6.1 shows an indent field of 10 by 10 indents in a pure iron particle of the SMCs. *The hardness of specimens with and without a deformation layer polished to step 7a and following procedure D, respectively, is shown in Figure 6.2 over the indentation depth of 900 nm. Both samples exhibited a pronounced indentation size effect [36]. At 900 nm indentation depth, a plateau value was almost reached for both samples. The hardness at 850 nm is 2.21 ± 0.21 GPa and 1.93 ± 0.14 GPa with and without a deformation layer, respectively (Table 6.1). At a larger depth, only a small effect of the deformation layer was observed, while at a lower indentation depth, the difference was quite pronounced. The hardness values of the two samples differed by more than 1 GPa at 250 nm indentation depth. Furthermore, the hardness of the specimen without a deformation layer showed a less pronounced depth dependence (Table 6.1). Young's modulus was independent of the indentation depth for both samples.* [41]

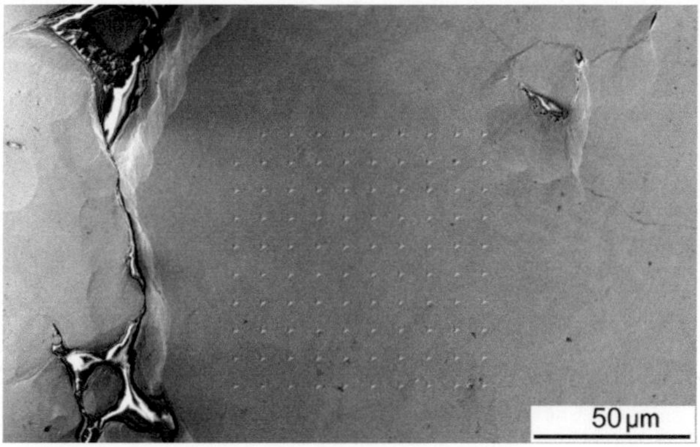

Figure 6.1: SEM micrograph of an indent field of 10 by 10 indents with an indentation depth of 300 nm in the pure iron particle of the SMCs.

Table 6.1: Average values of Young's modulus and hardness with their standard deviations measured at 250, 550 and 850 nm indentation depth for samples with different thicknesses of deformation layers. The thicker the deformation layer, the more the hardness measurements were affected at low indentation depths. Young's modulus was approximately the same for all measurements. [41]

Deformation layer thickness [nm]	Indentation Depth [nm]	Hardness [GPa]	Young's modulus [GPa]
–	250	2.21 ± 0.14	219 ± 8
	550	2.15 ± 0.15	229 ± 10
	850	1.93 ± 0.14	221 ± 11
163 nm	250	2.77 ± 0.12	197 ± 6
311 nm	250	3.25 ± 0.15	230 ± 8
	550	2.36 ± 0.18	212 ± 10
	850	2.21 ± 0.21	214 ± 11

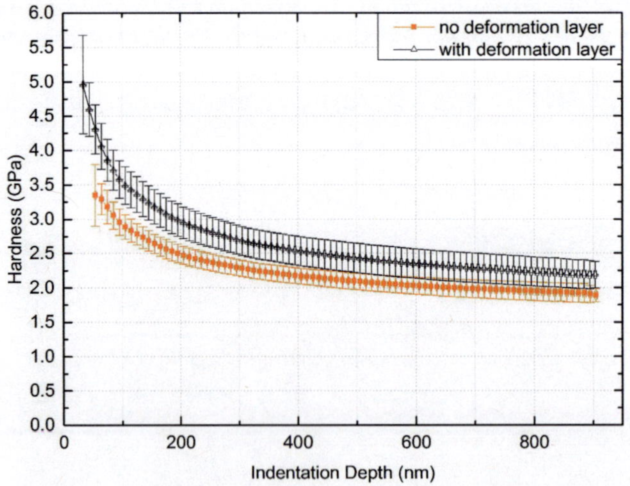

Figure 6.2: Hardness vs. indentation depth of SMC specimens with a ~240 nm thick deformation layer and without a deformation layer. Up to 900 nm indentation depth, both samples exhibited a pronounced indentation size effect and no plateau hardness is reached. The hardness of the sample with a deformation layer was higher, in particular for depths < 300 nm. The average value of 60 indents is shown with the error bars representing one standard deviation. [41]

6.2 Nanoindentation along a Particle Boundary

Indents were placed on a particle boundary of a specimen prepared following procedure B. Figure 6.3a shows the hardness and Young's modulus for the indents positioned along the boundary shown in Figure 6.3b. For comparison, the hardness and modulus values of a sample polished following procedure D (away from the particle boundaries) are indicated by the black and red lines, respectively. Not all indents were located exactly on the boundary, some of them indented the adjacent material of a particle. [41] Both hardness and Young's modulus are significantly lower for indents 8, 9 and 10 although Indent 9 was positioned exactly on the boundary (see Figure 6.4). No cracks were observed at the boundary indicating a somewhat ductile behaviour. Except for these three indents, there was a clear strengthening effect of particle boundaries. *For this particular boundary, the hardness was 3.45 ± 0.37 GPa, while for the indents close to the boundary, but in the adjacent particle, a hardness of 3.15 ± 0.42 GPa was determined. Compared to the bulk values of 2.77 ± 0.12 GPa of the iron particles (Table 6.1), the hardness values of the boundary region were clearly higher.* [41]

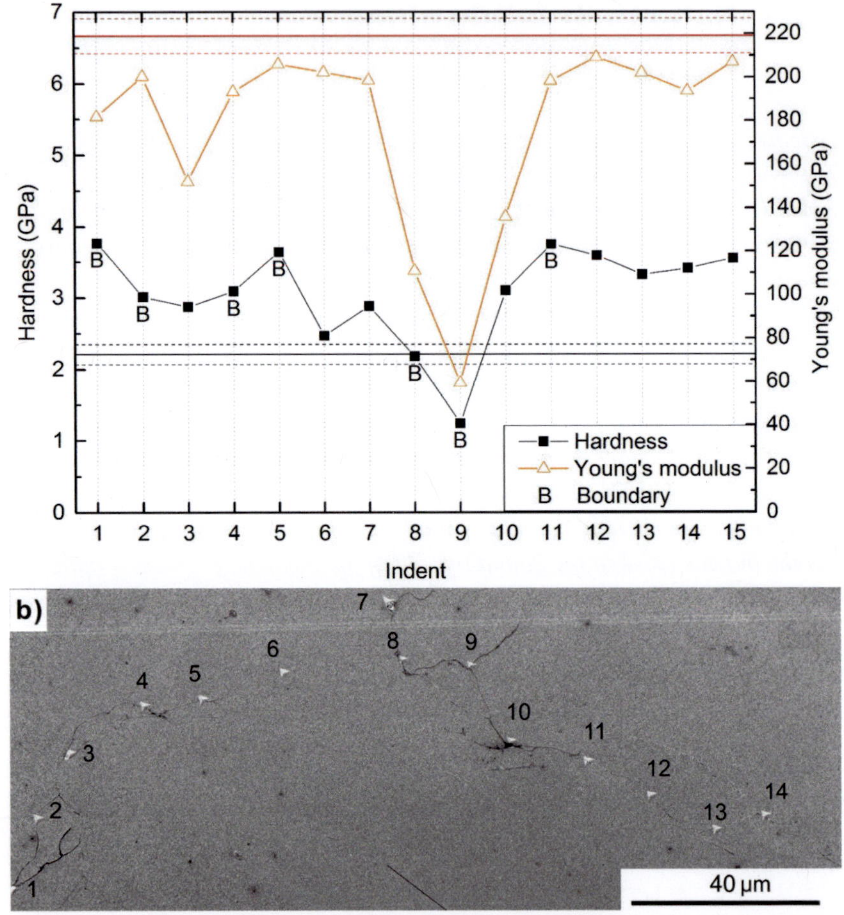

Figure 6.3: Nanoindentation of the boundary region: a) Hardness and modulus exhibited significant changes depending on the indent position. The influence of the subsurface and the deformation layer could be seen when compared to the hardness and Young's modulus of a specimen without a deformation layer (horizontal lines with dashed lines for the standard deviation). b) SEM of the boundary after indentation showed that not all of the indents were positioned directly on the boundary. [41]

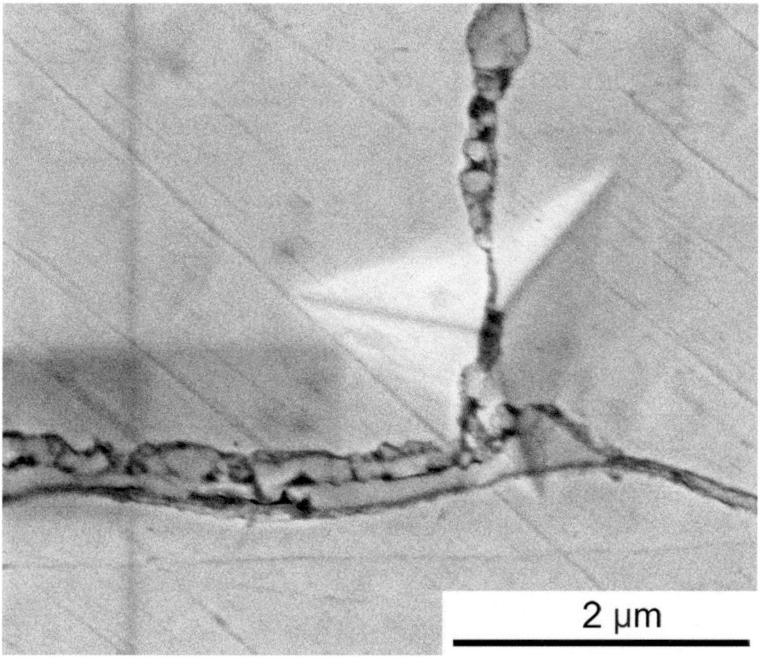

Figure 6.4: Indent 9 from the boundary tested in Figure 6.2. The indent was clearly placed on the boundary. [41]

6.3 Nanoindentation of Different Types of Particle Boundaries

Table 6.2 also summarises the hardness of each type of boundary measured on a fully oxidised specimen. The difference between the individual boundary types was not resolvable with nanoindentation, there were only minor differences that all lie within standard deviation. However, the particle boundaries were about 1.5 GPa harder than the pure iron particles. The hardness for indents close to the particle boundary in the adjacent iron particle was higher than for iron particles (see also section 6.2).

Table 6.2: Hardness of all types of particle boundaries in SMCs for a fully oxidised specimen. The boundaries are harder than the pure iron particles.

	Hardness [GPa]
type-1 boundary	3.41 ± 0.90
type-2 boundary	3.92 ± 0.70
type-3 boundary	3.97 ± 0.73
type-4 boundary	3.76 ± 0.90
particle (near boundary)	2.89 ± 0.53
particle (Table 6.1)	2.21 ± 0.14

6.4 Microcompression

Figure 6.5 shows engineering stress–strain curves from microcompression tests of pillars that were fabricated in the bulk or contain a particle boundary. *For the analysis, the compliance of the base material and the load frame [115] were accounted for.* [106]

Figure 6.5a show the engineering stress – strain curves for pillars fabricated in the bulk. *Strain bursts can be seen in the curves corresponding to clearly visible plastic slip lines in the pillars. Both single and multislip were observed as well as strain hardening with increasing strain.* [106] The multislip occurred either in one or two directions. The SEM-micrograph in Figure 6.4a shows multiple slip lines in one direction. The yield stresses lay between 313 MPa and 492 MPa.

Figure 6.5b shows the engineering stress-strain curves for pillars containing boundaries and exhibiting single slip together with a typical pillar after deformation. *The yield stresses were between 577 MPa and 671 MPa. These were the highest stresses both for pillars in the bulk and pillars with boundaries.* [106] The SEM-micrograph in Figure 6.5b shows a pillar containing a boundary and exhibiting a single strain burst.

Figure 6.5c shows the engineering stress-strain curves for pillars exhibiting multiple strain bursts. [106] The yield stresses tended to be lower than for pillars exhibiting only one strain burst and lay between 374 MPa and 632 MPa. It seems that pillars exhibiting multiple strain bursts in one direction showed higher yield stresses than pillars exhibiting multiple strain bursts in two directions. The SEM-micrograph in Figure 6.5c shows a pillar containing a boundary and exhibiting multiple strain bursts in two directions. The deformation behaviour of the investigated pillars containing boundaries appeared not to be influenced by them.

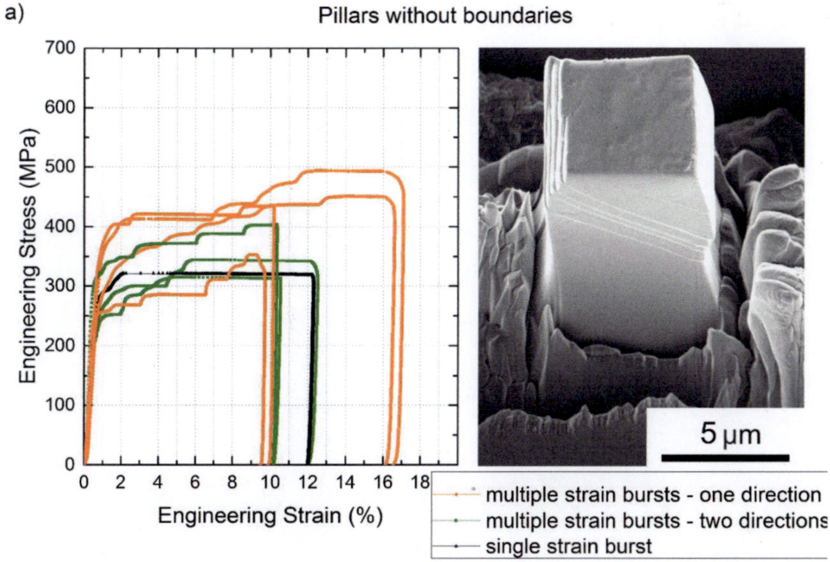

(Figure 6.5 continued on next page)

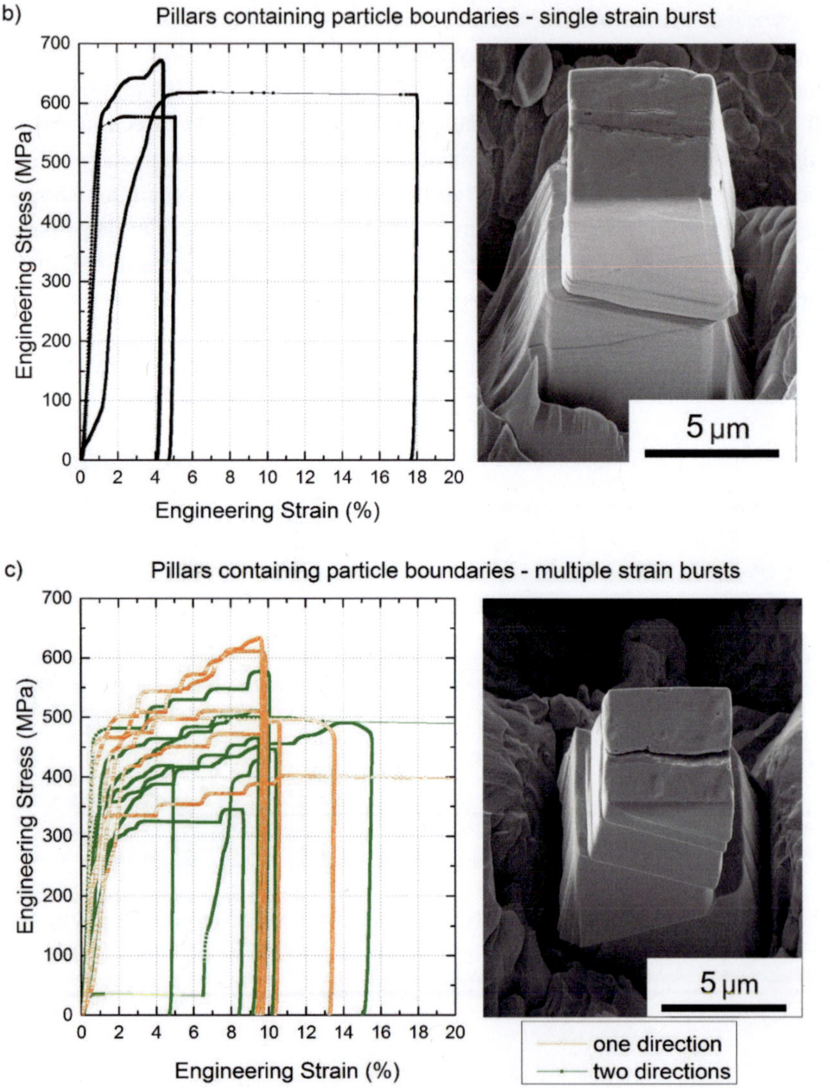

Figure 6.5: Engineering stress-strain curves and corresponding deformation characteristics of a) pillars in bulk particles (single and multislip), b) pillars containing particle boundaries with single strain bursts and single slip, c) pillars containing particle boundaries with multiple strain bursts and multislip. [106]

Considering the influence of pores on the mechanical behaviour of the SMCs, Figure 6.6a shows the engineering stress-strain curves of microcompression pillars containing particle boundaries and pores of different sizes. The pore size increased from Pillar 1 to 4. Pillars 1-3 containing different types of particle boundaries and small central pores did not influence the stress-strain behaviour and showed no deformation around the pores after compression (see Figure 6.6b). Figure 6.6c shows Pillar 3 after testing and although it is a larger pore, it did not show any signs of deformation or failure. Pillar 4 contained a pore filled with complex boundary layers consuming a large volume of the pillar. Figure 6.6c shows that Pillar 4 collapsed after compression resulting in a non-measurable engineering stress – engineering strain curve (Figure 6.6a).

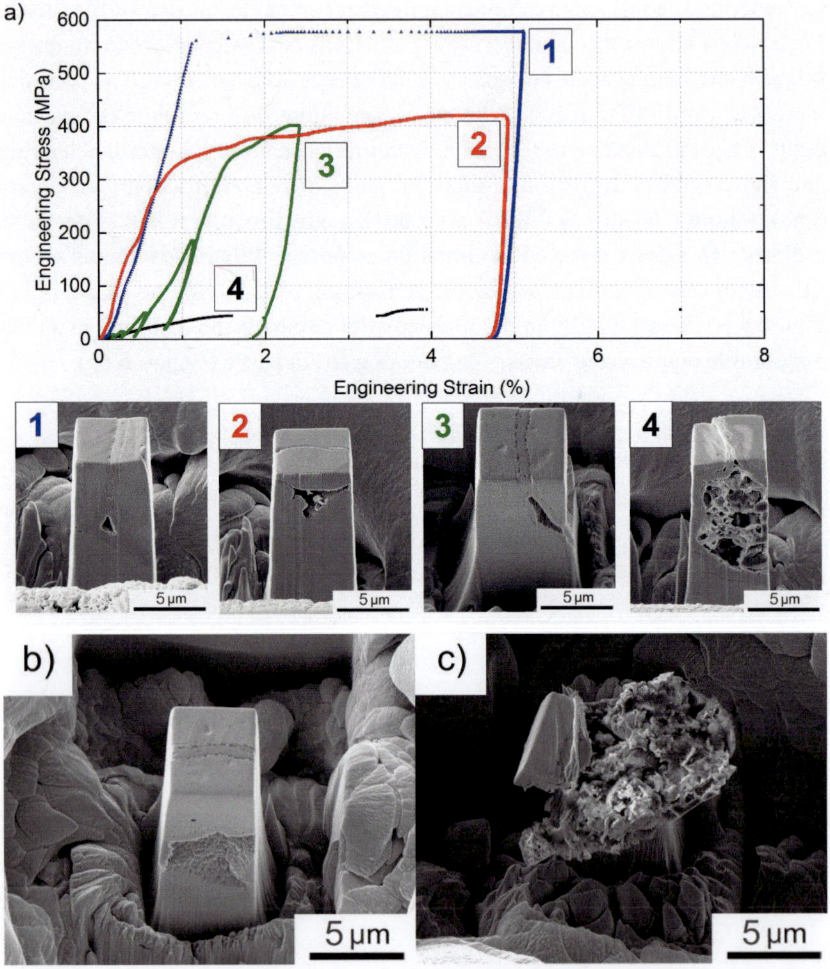

Figure 6.6: a) Engineering stress-strain of pillars containing particle boundaries and pores. The pore size increased from pillar 1 to 4, b) Pillar 3 after compression showing that the large central pore showed no deformation or failure, c) Pillar 4 collapsed during compression.

6.5 Microcantilever Bending

The loading curves of the microcantilevers shown in Figure 6.6 are sorted with respect to the particle boundaries, which determine the fracture behaviour of the cantilevers. [106]

Cantilevers containing type-1, -2, and -3 boundaries all failed in a brittle manner (Figure 6.7a) with little or no plastic deformation of the beams. Cantilevers with type-1 boundaries failed at the lowest loads, while those with type-2 boundaries sustained the highest loads of all brittle boundaries. [106] As shown in the SEM-micrographs in Figure 6.7a, cantilevers with type-1 boundaries exhibit brittle failure at the boundary. Cantilevers containing particle boundaries of type-2 and type-3 also failed in a brittle manner at the boundary.

Cantilevers with type-4 boundaries showed the highest plastic deformation (Figure 6.7b). The SEM micrographs in Figure 6.7b show that the cantilever exhibited ductile behaviour without fracture. The cross section (lower image in Figure 6.7b) shows the complex pore structure, which did not cause the formation of cracks or delamination.

The particle boundaries cannot always be assigned to one group only and may occur as mixed types [106]. In case of a type-3/type-4 combination, the plastic deformation was higher than the plastic deformation of cantilevers containing the brittle type-1 to -3 boundaries but lower than that of the cantilevers with ductile type-4 boundaries (Figure 6.7c). The SEM micrographs of Figure 6.7c show that the cantilever fails at the type-3 interface only, i.e. the interface between the phosphorous layer and the particle.

Figure 6.7d shows the load-displacement curves for cantilevers without particle boundaries. The loads were in the same range as for cantilevers containing boundaries. The plastic deformation, however, is lower than that for cantilevers with type-4 boundaries. The SEM-micrograph in Figure 6.7d shows that the cantilevers did not fracture.

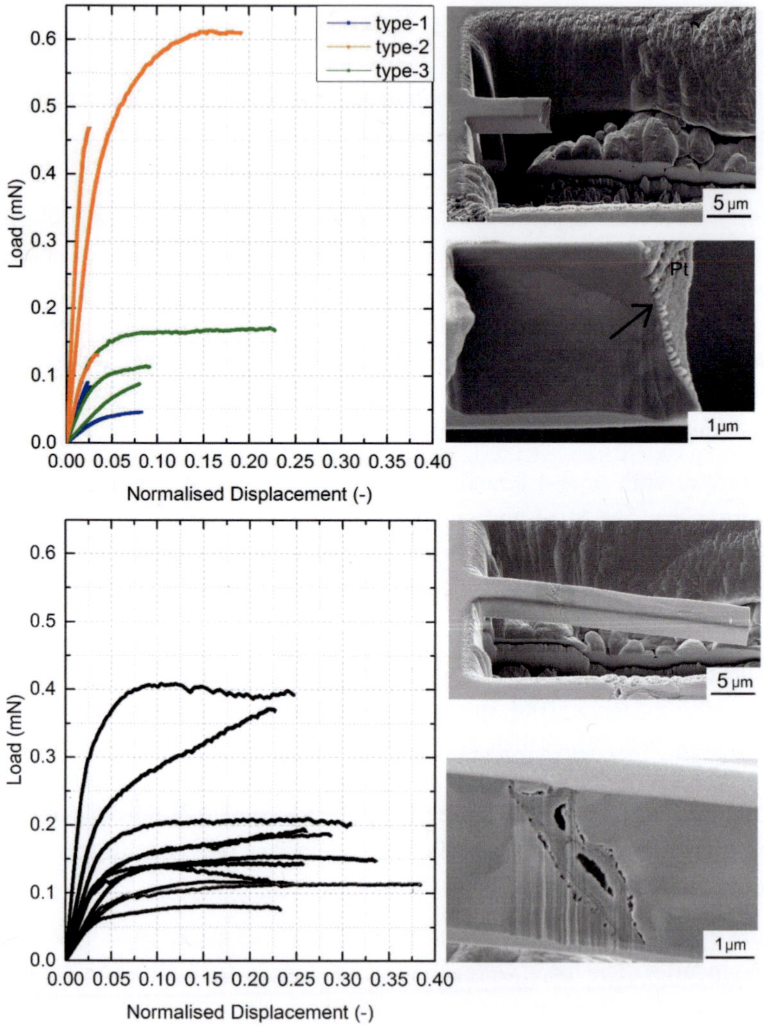

(Figure 6.7 continued on next page)

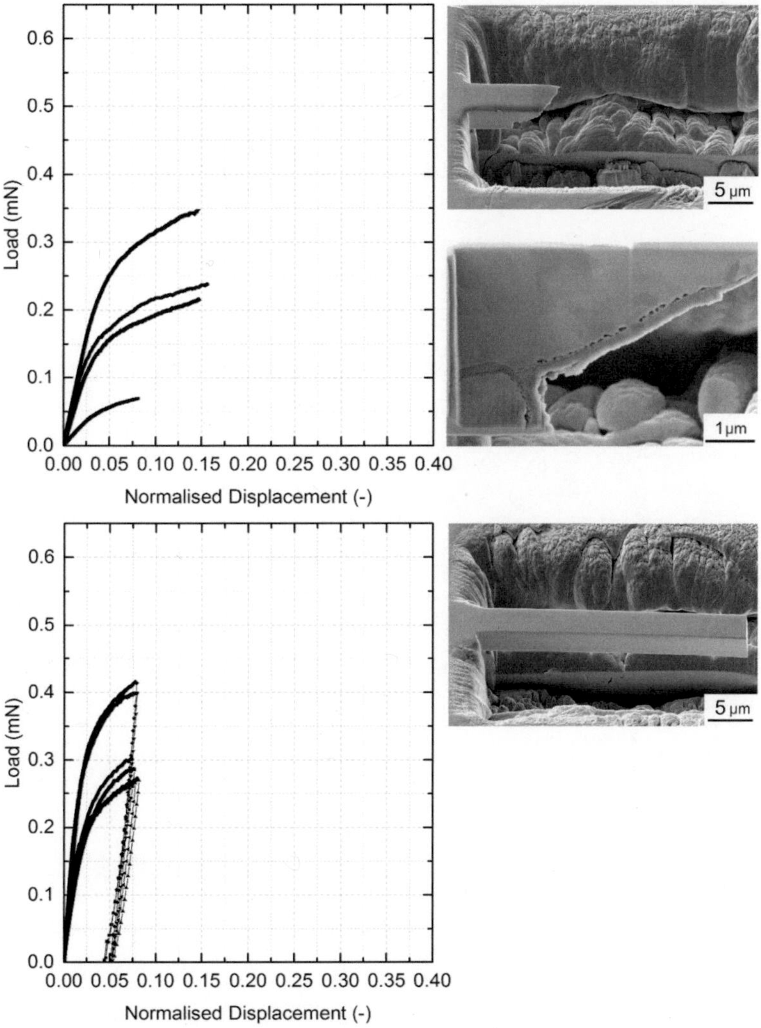

Figure 6.7: Load-normalised displacement curves and deformation characteristics for: a) cantilevers with type-1, type-2 and type-3 boundaries, b) cantilevers with type-4 boundaries, c) cantilevers with mixed type-3 and type-4 boundaries, d) cantilevers in the bulk without particle boundaries. The displacement was normalized by the respective lever length. [106]

6.6 Microtensile Testing

Figure 6.8a shows the stress-displacement curves of two microtensile specimens containing different types of particle boundaries which are shown in Figure 6.7b and c. The microtensile test specimen in Figure 6.8b contains a type-2 particle boundary lying in the centre of the specimen. The inset shows a pore at the end of the specimen. In Figure 6.8c, the specimen contains a type-3 boundary which also lies in the centre.

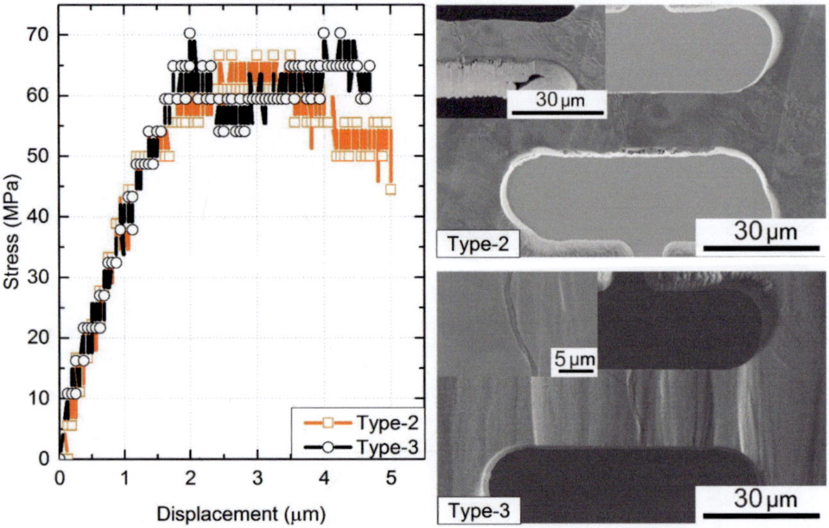

Figure 6.8: a) Stress-displacement diagram for microtensile test specimens with a type-3 (black) and type-2 boundary (orange). The type-3 boundary shows a stress drop after the maximum stress was reached. After the second maximum was reached, the specimen broke. The type-2 boundary shows only one maximum stress at which the specimen failed. b) Tensile test specimen containing a type-2 particle boundary in the centre and a pore at the end of the specimen, c) Tensile test specimen containing a type-3 boundary at the centre.

(microtensile test and SEM micrographs courtesy of Dr. Yoji Mine, Kumamoto University)

Both stress-displacement curves in Figure 6.8a show a good agreement in the elastic regime. The spiky curve propagation is due to the difference in resolution of the recording devices for the displacement and stress during testing. The black curve corresponds to a tensile test specimen with a type-3 boundary. The stress increases linearly up to a maximum stress of 70.3 MPa, after which it drops to 59.5 MPa, then reaches a second maximum of 70.3 MPa and finally the specimen fails. The curve propagation is correlated to the video that was taken during testing. The video helped to understand the crack propagation but the image sequence of the video is not shown here. At the first maximum, the crack opened. After the initial crack opening, the stress dropped and there was only minor further crack opening until the second maximum stress of 70.3 MPa at which the final failure occurred and the specimen broke.

The orange curve in Figure 6.8a corresponds to the stress-displacement behaviour of a specimen containing a type-2 boundary. After the linear increase of stress, plastic deformation starts, a maximum stress of 66.6 MPa is reached and the stress decreases again up to the final failure without any stress drops. In this case, the video did not show any crack opening.

Figure 6.9a shows an SEM micrograph of the specimen containing a type-2 boundary after testing. It clearly did not fail at the expected boundary but at a type-1 boundary connected to a pore near the end of the prepared specimen. The type-2 boundary which was supposed to be tested remained unaffected and does not show any signs of cracks. The curtaining at the sides of the specimen did not influence the testing. The fracture surface appears smooth indicating a brittle fracture. Figure 6.9b shows an SEM micrograph of the specimen containing a type-3 boundary after fracture. The specimen broke as expected at the boundary and shows a smooth fracture surface. The lines on the sample surface are preparation artefacts from the cross-sectional ion beam polishing.

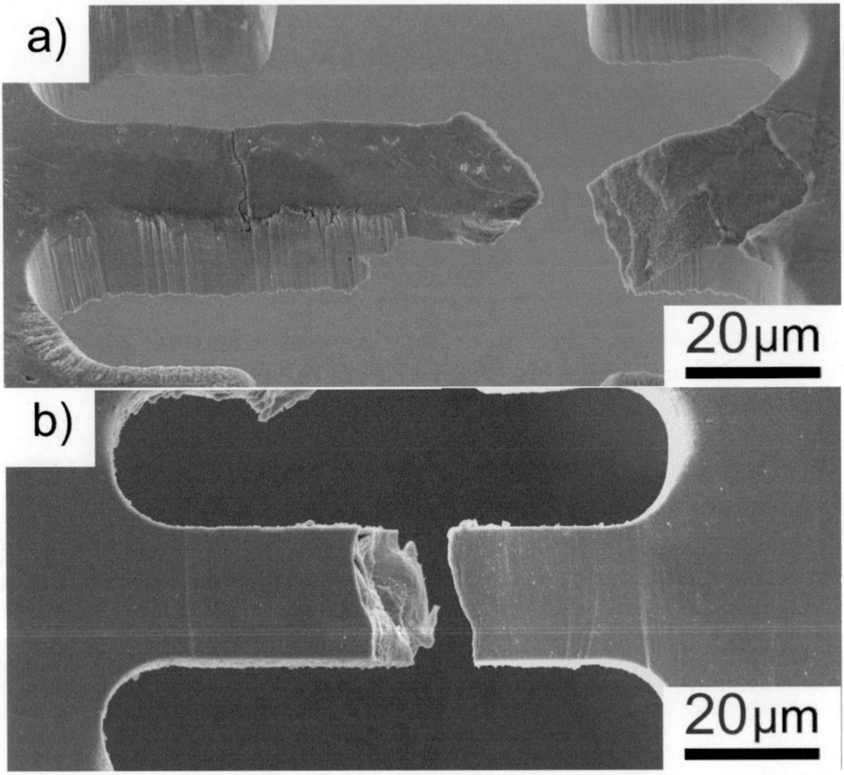

Figure 6.9: a) Specimen containing a type-2 boundary after testing. It did not break at the expected type-2 boundary but on a type-1 boundary containing a pore, b) Specimen containing a type-3 boundary after testing. The failure occurred in the boundary.

(SEM micrographs courtesy of Dr. Yoji Mine, Kumamoto University)

7 Macroscopic Fracture Toughness Testing

Figure 7.1a shows a light microscopy image of the crack propagation of a partially oxidised specimen A after four point bending. The main crack started near the notch tip at a boundary. Smaller cracks also started at the left and right faces of the notch. The main crack shows branching and did not break the specimen. Some of the specimens showed very long running branches. There was no visible difference between cracks of specimens A and B. The SEM micrograph in Figure 7.1b shows the surface of the crack at its opening. The surface appears smooth as in a brittle failure. More brittle fracture surfaces are observed for type-1 to type-3 boundaries (Figure 7.1c and d).

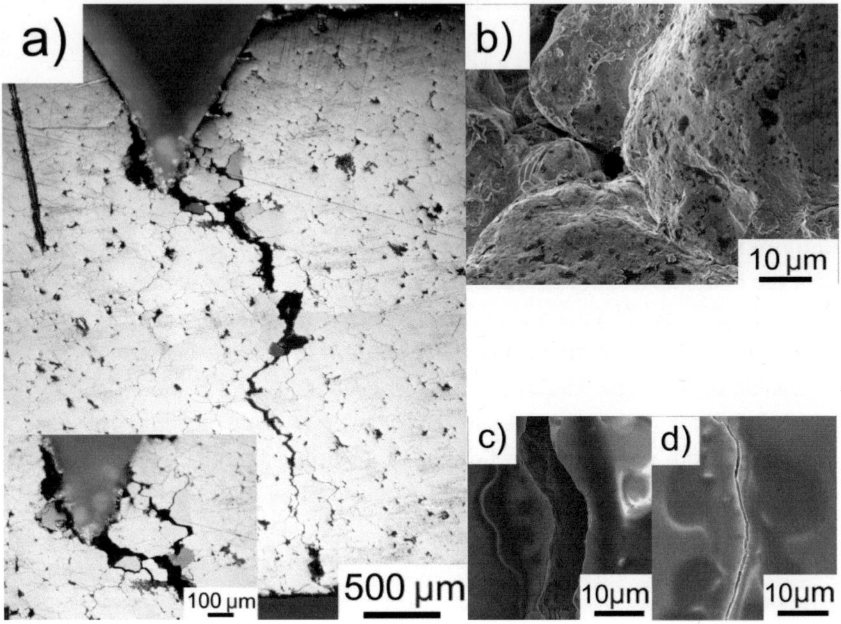

Figure 7.1: Fractographic analysis of a partially oxidised SMC after four point bending testing. a) Light microscopy image of the crack in a specimen A, b) SEM micrograph of the smooth fracture surface at the crack opening of a partially oxidised SMC indicating brittle fracture, c) SEM micrograph of a partially opened brittle crack d) SEM micrographs of a partially opened crack of a crack branch.

Figure 7.2a shows a crack of a fully oxidised sample. There is no difference compared to the partially oxidised specimens in the light microscope. The cracks also show branching. For fully oxidised SMC, a further type of fracture surface is observed due to the higher type-4 boundary content. The fracture surfacc of a typc-4 boundary surfacc appcars rough in thc SEM which comes from the "left overs" of bridges from type-4 boundaries (see Figure 7.2b). Bridges form between the crack surfaces as seen in half opened cracks (Figure 7.2c). Even in fully opened cracks the leftovers from the bridges were found (Figure 7.2d).

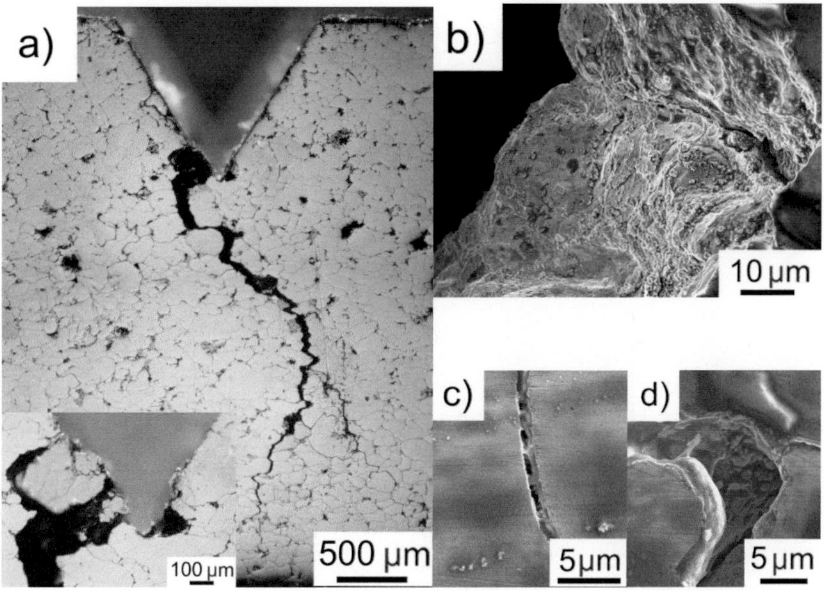

Figure 7.2: Fractographic analysis of a fully oxidised SMC after four point bending testing. a) Light microscopy image of the crack of a fully oxidised specimen, b) SEM micrograph of the rough fracture surface at the crack opening, c) SEM micrograph of a partially opened crack showing bridging of a type-4 boundary, d) SEM micrograph of a left-over from crack bridging of a type-4 boundary in a opened crack.

Figure 7.3 shows the force-displacement curves of fully and partially oxidised specimens. The mean maximum forces for specimens A and B were 248 ± 31 N and 236 ± 18 N respectively showing that there is no difference between them. However, partially oxidized specimens endure less strength than fully oxidized specimen with a mean maximum force of 288 ± 23 N. The discontinuity in the curves only occurred during the displacement-controlled experiments and is also not rate-dependent. The videos do not show a correlation of the maximum force and the crack opening at the observed surface, i.e. the crack opening of either the main crack or a side branch sometimes occurred at a lower force.

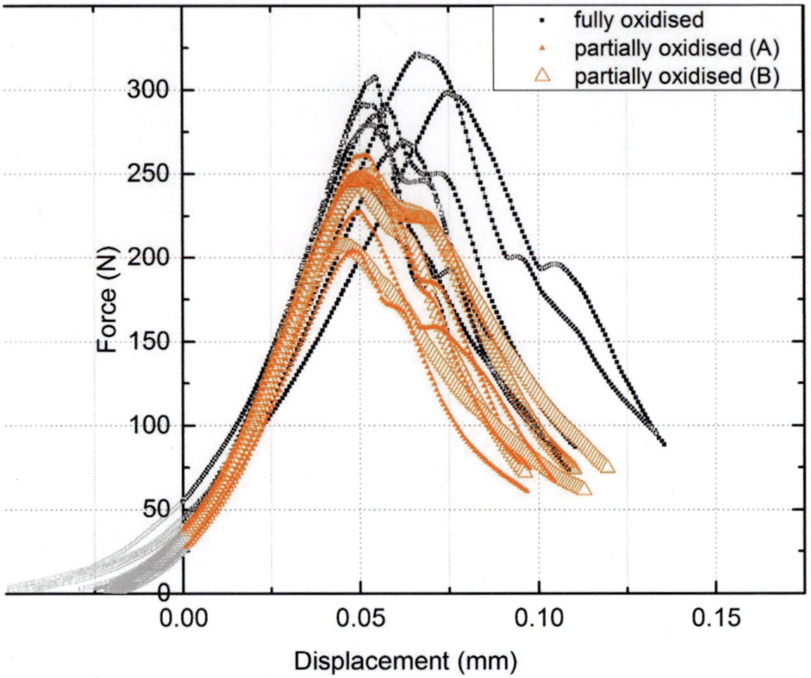

Figure 7.3: Force-displacement diagrams for fully oxidised specimens and partially oxidised specimens A and B. Fully oxidised specimens reach the highest forces. Specimens A and B show the same range of forces

8 Discussion

This research work has two main parts: The microstructure of Somaloy 700 3P was characterised and its mechanical properties on the micro- and macroscale were analysed. The used micromechanical testing methods were then critically assessed in terms of sample preparation and the testing methods themselves.

8.1 Deformation and Fracture Properties of Somaloy 700 3P at Different Scales

The main issues to be discussed are the microstructural characterisation and deformation behaviour of the pure iron particles on the microscale, the microstructural characterisation of the particle boundaries and their role in deformation and fracture, and the influence of the pores on the mechanical performance of the SMC.

8.1.1 Deformation Behaviour of the Pure Iron Particles in Somaloy 700 3P

For a full characterisation of Somaloy 700 3P, its initial condition as a powder in its unprocessed state was analysed. Somaloy 500 and Somaloy 700 3P have different particle size distributions and different coatings. The particle size distribution for Somaloy 700 3P is given in Table 3.1 with 68 % of the particles being between 150-425 μm in diameter. The mean particle size is larger than that of Somaloy 500 with 100 μm [12, 27]. The SEM micrographs shown in section 5.1 confirm the different particle sizes and also show water-atomised surface and the additional binder layer of the Somaloy 700 3P particles. The phosphorous coating of both Somaloy 500 and Somaloy 700 3P is not resolvable in the SEM.

The particles of Somaloy 700 3P were analysed after compaction and a heat treatment at 530°C under water vapour by TEM, SEM and EBSD revealing two main characteristics of the pure iron particles: i) the particles are not necessarily single-crystalline, and ii) they are heavily deformed in the compaction process (see section 5.2 and 5.3). IPF and IQ maps with a step size of 0.2 µm revealed that the iron particles are polycrystalline with a few grains per particle (see Figure 5.3). The EBSD maps also show colour variations within the particles indicating a subgrain structure. The subgrains containing a high dislocation density were also observed in TEM and SEM (see section 5.3). The iron particles undergo a strong cold-working during the compression at 800 MPa. It was reported that recovery and recrystallization occurs during annealing which partly or fully reduces the cold-working depending on the annealing parameters [17, 33]. At an annealing temperature of 530°C, the iron particles still show internal stresses whereas a temperature of 700°C leads to a stress relaxation of the material [33]. In terms of the magnetic properties, the hysteresis loss is increased by the cold-working because the higher dislocation density hinders the movement of the domain walls [17, 18]. It is postulated in [33], though, that the high dislocation concentration in the pure iron particles is not the only cause for the hysteresis losses but that the particle boundaries strongly influence the movement of the Bloch walls. As the permeability depends on the microstructure, it might also be an influenced by the particle boundaries.

In terms of mechanical properties, Young's modulus of the pure iron particles of Somaloy 700 3P was found to be independent of the indentation depth for all samples. The determined mean value of Young's modulus was 217 ± 11 GPa averaged for all results given in Table 6.1 and the hardness measured was 2.21 ± 0.14 GPa. The mechanical properties determined in this study appear reasonable, while a direct comparison with literature data is difficult. The hardness and Young's modulus of the bulk iron particles was determined by nanoindentation (see section 6.1). Indentation fields of 10 by 10 indents were placed inside the iron particles with an indentation depth of 300, 600 and 900 nm, respectively. A Young's modulus for the bulk iron particles was not reported in literature, however, for pure polycrystalline iron it is about 210 GPa [116]. Values for the SMC as such are given in literature stating a Young's modulus of 190 GPa for Somaloy 700 3P (see Table 3.2) [95] and a modulus of about 130 GPa for Somaloy 700 annealed in air [117]. Indentation of electropolished Somaloy 500 yielded microhardness values of 1.6 GPa [17], which is not much lower than the hardness of 1.93 ± 0.14 GPa at 900 nm indentation depth found in this work (Table 6.1). The hardness of a Somaloy sample mechanically polished with colloidal silica as the final step (particle size of 0.04 μm) was reported in the range of 2 GPa at depths < 300 nm [18]. As one expects a clear effect of the surface preparation method and the indentation size effect, the agreement with the findings is reasonable.

The microcompression pillars fabricated in the bulk iron particles of Somaloy 700 3P exhibited more strain bursts, more pronounced hardening as well as higher flow stresses compared to previous investigations of single-crystalline pure α-Fe pillars of the same size (see Figure 8.1) [118]. The pillars with a top side length of 5 μm were milled in bulk iron particles of Somaloy 700 3P and were likely single-crystalline due to their size. Figure 8.1 shows the engineering stress – engineering strain diagram for pure α-Fe and the pure iron particles of Somaloy 700 3P. Pillars of 4.7 μm exhibit a yield stress of about 200 MPa, whereas the pillars of this work being only a bit larger show yield stresses at 5 % plastic strain of about 300-450 MPa. These values are similar to the ones of pure α-Fe pillars with a diameter of 2.6 μm. The cold-working of the particles may explain the higher stresses observed in this work.

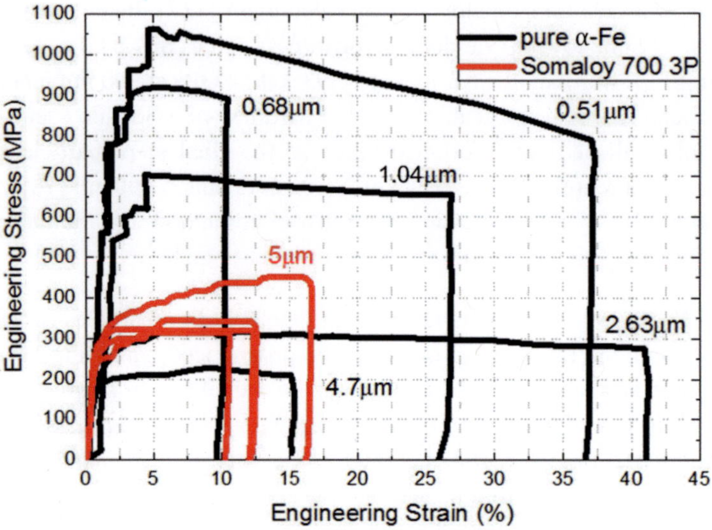

Figure 8.1: Engineering stress vs. engineering strain for different sizes of α-Fe single-crystalline pillars from [118] and 5 μm presumably single-crystalline pure iron pillars made from Somaloy 700 3P which show higher stresses and more strain bursts than α-Fe of the same size.

In Figure 8.2, the yield stress at 5% plastic strain in microcompression pillars in the bulk is compared with the maximum yield stress in the microcantilever tests. Pillars containing no particle boundaries do not show a pronounced difference in yield strength for single and multiple strain bursts. The strength estimate of the cantilevers is higher than in microcompression, which is likely related to the analysis method. The maximum yield stress was estimated taking the maximum load in the load-deflection curves of the beams that deformed plastically [67]. Since the maximum load might be higher than the load needed for 5% plastic strain, the resulting estimate of the yield stress might be also too high. However, the stresses in the microcantilevers are almost twice as high as in the pillars which may not be caused only by the different methods used for calculating the stress. It can furthermore be argued that the effectively deformed volume of the pillars is smaller than the volume of the microcantilevers which show a strain gradient as stated in [119].

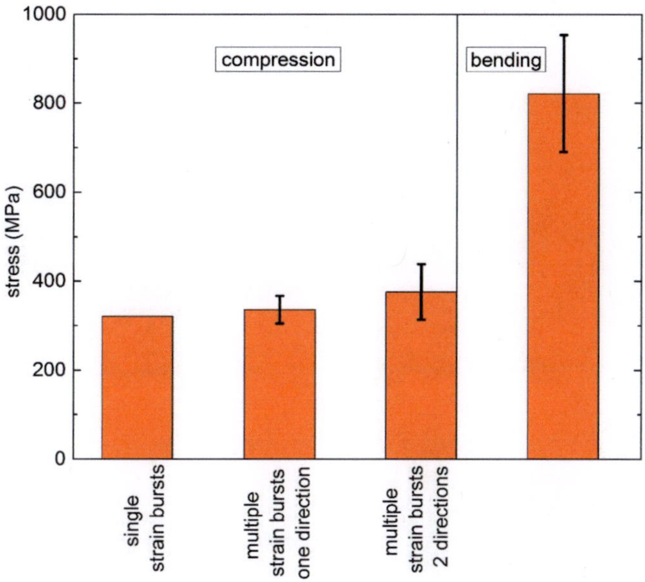

Figure 8.2: Yield stress at 5% plastic strain for eight tested pillars in the bulk iron particles of Somaloy 700 3P and at maximum load for six cantilevers in the bulk. [106]

8.1.2 The Role of the Particle Boundaries in the Deformation and Fracture Mechanisms of Somaloy 700 3P

A new model for describing the complex particle boundaries of Somaloy 700 3P heat treated at an annealing temperature of 530°C was found based on the TEM micrographs in section 5.3. In [33], a model was given for Somaloy 700HR-3P stating that an iron oxide layer forms between the iron particles and the phosphorous layer or additionally within the phosphorous layer (see Figure 1.6). The new model classifies the particle boundaries into four main types depending on the complexity of layers (see Figure 8.3).

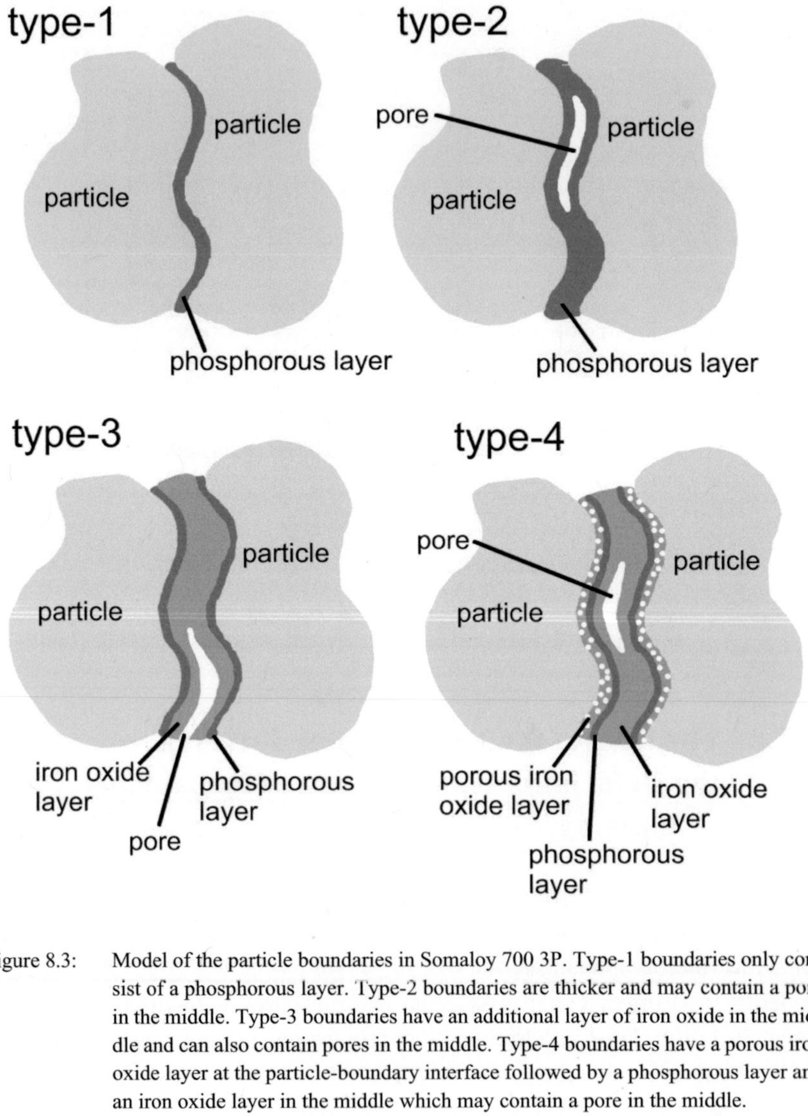

Figure 8.3: Model of the particle boundaries in Somaloy 700 3P. Type-1 boundaries only consist of a phosphorous layer. Type-2 boundaries are thicker and may contain a pore in the middle. Type-3 boundaries have an additional layer of iron oxide in the middle and can also contain pores in the middle. Type-4 boundaries have a porous iron oxide layer at the particle-boundary interface followed by a phosphorous layer and an iron oxide layer in the middle which may contain a pore in the middle.

The type-1 boundary is comparable in composition and thickness to the boundary found in Somaloy 500 [28, 29]. Except for the type-1 boundary, all other types may contain pores in the middle. Mixed states of particle boundaries may also occur, in particular mixed type-3/type-4 boundaries. It is expected that type-1 and type-2 boundaries behave similarly because both only consist of a phosphorous layer. Type-3 and type-4 boundaries are more complex and should, thus, behave differently. Here, a "boundary" addresses all layers and their corresponding interfaces. If only a particular layer or interface is considered, it is named accordingly. In literature, the surface of Somaloy 500 particles and the particle boundaries of Somaloy 700 were examined by EDX and XPS finding oxygen, phosphorus and iron as main chemical elements in the SMCs [28, 33]. These results can be confirmed by the EDX measurements in the TEM in this research work, although the exact chemical composition of the boundary layers remains unknown. The thicknesses of the individual layers and the amount of iron, oxygen and phosphorus could be determined (see section 5.3). From the channelling contrast in both TEM and SEM micrographs, it can be assumed that the porous iron oxide layer of the type-4 boundary has a different chemical composition than the iron oxide layer in the middle of the boundary.

There are two important properties of the particle boundary that were revealed by nanoindentation. First, the particle boundaries strengthen the SMC. Second, they do not exhibit cracks after indentation. [41] It can be argued that the particle boundaries themselves are hard but also not very brittle. Using nanoindentation, indents were placed along the particle boundaries with an indentation depth of 300 nm. The indents were either placed directly on or near the boundary. The particle boundaries strengthen the SMC as described in section 6.2 and 6.3. *One could even interpret a gradual decrease of the hardness with increasing distance from the boundary, when comparing the hardness of the boundary and the material next to it. However, the difference is admittedly small with standard deviations greater than 10%.* [41] The particle boundaries are about 1.5 GPa harder than the inside of the pure iron particles. The different types of particle boundaries show only minor changes in hardness with a large scatter. *The layers, the particle boundaries consist of, are usually assumed to be brittle, and cracks would, thus, be expected after indentation, either at the surface or underneath the indent (see Figure 8.4)* [41]. The Berkovich indenter leaves an imprint in the shape of an equilateral triangle with a lateral size of 2.3 μm on the surface after indenting to a depth of 300 nm. Because the size of the imprint is larger than the thickness of the particle boundaries, it was neither possible to create a crack in the particle-boundary interface nor to measure boundary type-specific properties alone.

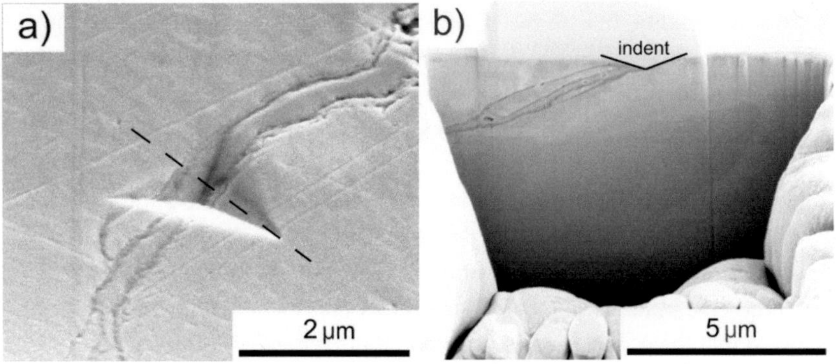

Figure 8.4: a) Indent at 300 nm indentation depth in a type-4 boundary, b) Cross section through the indent. Although the boundary was indented, there are no cracks in the boundary layers.

In order to resolve the mechanical properties of the individual types of boundaries, further testing methods were used i.e. microcompression, microcantilever bending and microtensile testing. *Microcompression testing showed that neither the type of boundary nor their volume fractions in the pillar appear to influence the deformation mechanism or the yield stress significantly. The particle boundaries appear to strengthen the pillars even in the presence of pores.* [106] Pillars were fabricated from Somaloy 700 3P containing an individual boundary. In Figure 6.5, the engineering stress – engineering strain curves for pillars containing particle boundaries are classified according to their deformation behaviours. Pillars with single strain bursts show higher stresses than pillars with multiple strain bursts. The stresses are also higher compared to pillars containing no boundaries. Figure 8.5 shows the yield stresses at 5% plastic strain for pillars with and without particle boundaries. In comparison to the pillars in the bulk, which show similar stresses for the different types of strain bursts, the pillars containing particle boundaries show different stresses. The pillars exhibiting single strain bursts show the highest stresses. Pillars exhibiting multiple strain bursts in one or two directions tend to have lower stresses. The particle boundaries were observed to further increase the stresses compared to pillars made inside the particles. In macroscopic test, high yield stresses are also observed [35]. While the pillars showed plastic deformation and surface cracks (see Figure 8.6), failure was never observed along the particle boundaries.

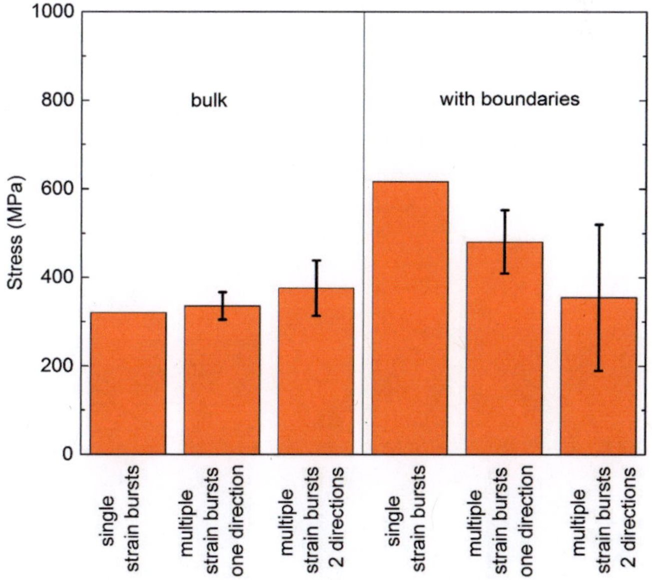

Figure 8.5: Yield stress at 5% plastic strain for eight pillars in the bulk and 18 pillars containing particle boundaries. [106]

The new model of the complex types of particle boundaries cis verified by analysing the SEM micrographs of the pillar surfaces after compression. Cracks were observed for all type-1 to type-3 boundaries on the top surface of the pillars at the particle-boundary interface. There is a mixed failure observed for mixed type-3/type-4 boundaries failing at the type-3 particle-boundary interface (see Figure 8.6b). Type-4 boundaries do not show cracks at all. Figure 8.6a shows the only case in which a complete fracture of a boundary during compression was observed. The top part of the pillar was separated during compression. In Figure 8.6b, the crack occurred at the particle-boundary interface or to be more precise at the particle-phosphorous layer interface of the type-3 boundary. Figure 8.6d shows the cross section though a pillar containing a type-2 boundary (Figure 8.6c). The crack appears only at the surface of the pillar. The boundary was only in the top third of the pillar and contained a small pore. There was also a pore in the bulk particle. The slip had occurred below the boundary in a single-crystalline area. *The particle boundaries may also occur as mixed types; on the top surface of the pillar, cracks along the phosphorous layer of the type-3 part (Figure 8.6e) of a mixed type-3 / type-4 boundary can be seen. The cross section through the pillar, revealed a thick type-4 boundary, and slip had occurred in the single-crystalline particle underneath the boundary (Figure 8.6f). No cracks or other signs of failure can be seen in the boundary indicating an improved strength of type-4 boundaries.* [106]

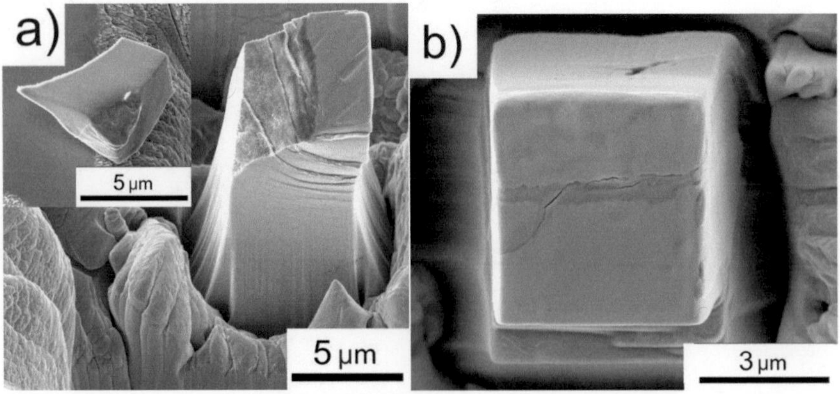

(Figure 8.6 continued on next page)

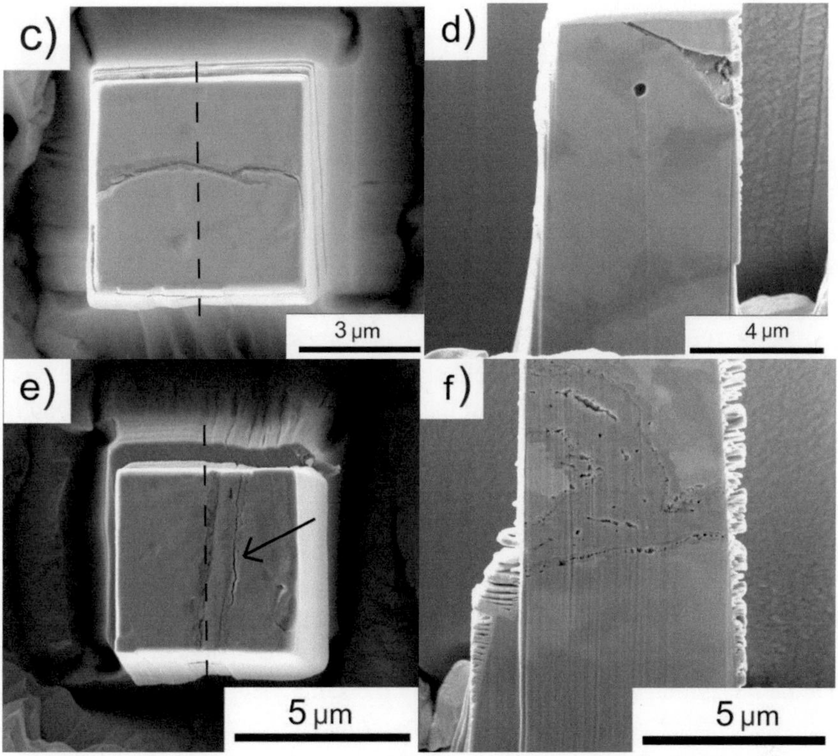

Figure 8.6 a) Pillar containing a type-1 boundary. A part of the top of the pillar broke off during compression, b) A large type-2 boundary cracked during compression, c) Pillar containing a type-3 boundary which cracked during compression. d) The cross section through the pillar in c) shows a small pore and a pore within the particle. e) The top surface shows a mixed type-3 / type-4 boundary, which shows cracks along the phosphorous layer. f) The cross section through the pillar in e), however, revealed a type-4 boundary but no cracks after deformation. (Figure 8.6e and f are from [106])

Overall, the findings of the microcompression tests are in agreement with the nanoindentation data. The boundaries lead to a local strengthening of the iron particles. Also, the boundaries are not very brittle and do not promote catastrophic failure by fraction at least under somewhat favourable loading conditions.

In bending, the particle boundaries weaken the microcantilevers. While cantilevers containing type-1 to type-3 boundaries break in a brittle manner, cantilevers with type-4 boundaries show a somewhat ductile behaviour. The loading curves in Figure 6.7 are sorted with respect to the type of boundary. Cantilevers with type-1 boundaries failed at the lowest loads, while cantilevers containing type-2 boundaries exhibit the highest loads. Type-1 to -3 boundaries failed in a brittle manner whereas mixed type-3/type-4 boundaries failed only at the brittle type-3 boundary. Failure occurred always along the particle-boundary interface, i.e. between the phosphorous layer and the iron particle. Cantilevers containing type-4 boundaries did not fail and, thus, show a somewhat ductile behaviour. The corresponding yield stress of the tested microcantilevers was calculated by taking the maximum load in the load-deflection curves. Due to the varying dimensions of the cantilevers, particularly in the width and thickness, the highest loads do not necessarily lead to the highest stresses. Figure 8.7 shows that cantilevers containing no particle boundaries reach the highest yield stress, while all types of particle boundaries appear to cause lower values. Type-1 boundaries resulted in the lowest yield stress, while type-2, type-3 and the mixed type-3/type-4 boundaries gave all about the same values. Cantilever beams with type-4 boundaries, which are the only ones leading to a pronounced ductile behaviour, exhibit approximately the same yield stress as the cantilevers with brittle particle boundaries. Comparing these results to the microcompression testing shows a contradictory behaviour between compression and bending. In microcompression, the deformation of the particle determines the failure strength whereas in bending, the particle boundaries limit the strength. *However, the stress states of the two test methods are different. Bending leads to a complex, multiaxial stress state with the maximum tensile stresses at the surface of the micro cantilever close to the fixed end where the particle boundaries are located. This basically results in tensile loading of the particle boundaries and failure of types 1-3.* [106] Again, for type-4 particle boundaries, no fracture was observed but deformation occurs at somewhat lower stresses than for cantilevers without a boundary. It may be argued that this is because the boundary helps to accommodate the strain gradient and, thus, reduces its effect on the measured strength. *Furthermore, bending leads to a multiaxial stress state with the maximum tensile stresses at the surface of the microcantilever close to the fixed end where the boundaries are located.* [106] This basically results in tensile loading of the boundaries and thus reduces the overall strength of the boundaries.

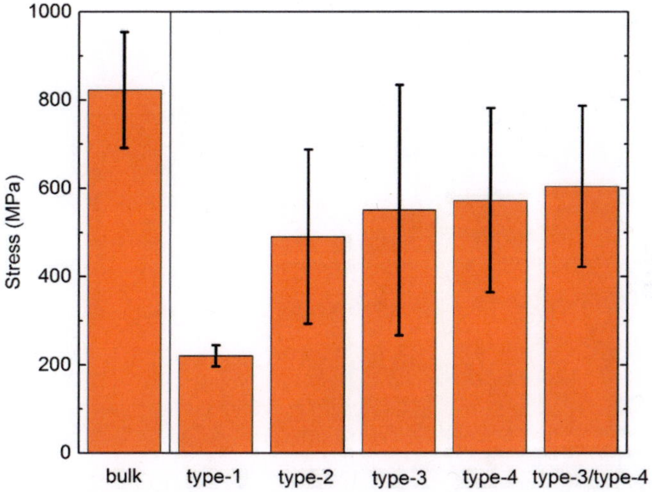

Figure 8.7: Yield strength at maximum load for six cantilevers in the bulk and 24 cantilevers containing particle boundaries. The four different types of particle boundaries and a mixed type-3/type-4 boundary are plotted. [106]

With the results from microtensile testing, it can be confirmed that the type-1 to 3 boundaries break in a brittle manner. To be more precise, the particle-boundary interface determines the failure. The type-4 boundary shows a ductile behaviour correlated to the porous iron oxide layer. The fracture surfaces of the type-1 and type-3 boundaries in microtensile testing (see Figure 8.8) as well as the fracture surfaces of single tested type-1 and type-3 boundaries in microbending (see Figure 6.6) show a smooth morphology indicating a brittle fracture behaviour. In microcompression, cracks were observed only in the brittle type-1 to -3 boundaries as well as in the type-3 part of a mixed type-3/4 boundary. The type-1 boundary in Figure 8.8b shows a very smooth fracture surface whereas the boundary in the pore is rough and of a different type.

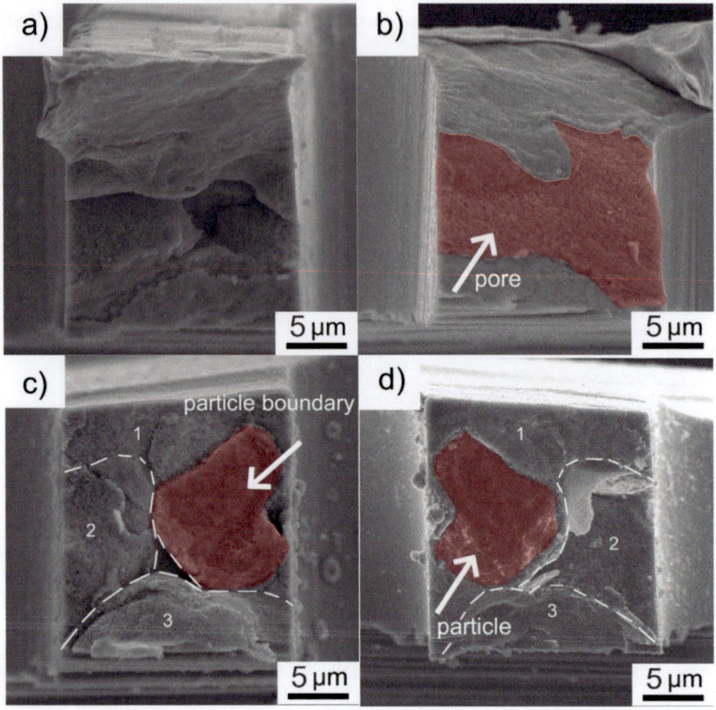

Figure 8.8: a) Fracture surface of a type-1 boundary with a smooth structure, b) Opposite fracture surface of a) showing that a pore with a rough surface (marked in red), c) Fracture surface of a type-3 boundary showing 3 particles. The crack ran first along the particle-boundary interface, particularly on particle 3. Particle 1 and 2, though, are only partially covered by the boundary (marked in red on particle 1), d) Opposite fracture surface of c). The crack crossed the particle boundary in particle 1 and 2 ran along the other particle-boundary interface (the uncovered particle 1 is marked in red).

(SEM micrographs courtesy of Dr. Yoji Mine, Kumamoto University)

As the type-1 boundaries are only a few tens of nanometres in thickness, it is hard to tell whether the fracture occurred in the boundary or at the particle-boundary interface. The fracture surfaces in Figure 8.8c and d belong to the same specimen containing a type-3 boundary and show three particles. In Figure 8.8c, particle 1 and 2 are partially covered by a particle boundary (marked in red in particle 1), whereas particle 3 is fully covered with the particle boundary. In Figure 8.8d, the opposite is the case (the uncovered particle 1 is marked in red). This indicates that the crack first ran along the particle-boundary interface, particularly on particle 3. In particle 1 and 2, though, the crack crossed the boundary, ran along the other particle-boundary interface dividing the particle boundary in two parts. This shows that the ductile boundary layers can also crack.

The tension-compression asymmetry observed at the macroscale with stresses approximately three times higher in compression than in tension [35] is also present at the microscale. In compression loading at the microscale, the failure strength is dominated by the strength of the iron particles while in bending, the tensile stresses at the upper side of the cantilever determine the failure in the boundary. In the microtensile tests from section 6.6, specimens containing a type-2 and a type-3 boundary were tested. Similar stresses of 66.6 MPa and 70.3 MPa are obtained for the specimens containing a type-2 and type-3 boundary, respectively. The microtensile test strengths correspond to the macroscopic tensile test strength of ~ 60 MPa [35]. The strength in tension is approximately seven times lower than in compression at the microscale.

Due to the different mechanical behaviour of the types of particle boundaries, the distribution of the particle boundaries in the SMC seems to be critical. In macroscopic four-point bending, higher forces are obtained in fully oxidised SMC compared to partially oxidised SMC. Moreover, Somaloy 500, reaches only strength values about a magnitude lower. In chapter 7, the load-displacement curves of partially of fully oxidised SMC were determined according to the SEVNB method. Specimens A and B covering a transition region with gradual increase or decrease of oxidation are very similar which explains the identical crack propagation and the similar mean forces of the partially oxidised SMC. The relatively high scatter of the force-displacement curves shown in Figure 7.3 correlates to the heterogeneity of the material and may be reduced by a larger set of experiments for a more quantitative analysis. The fully oxidised SMC reached higher loads than the partially oxidised SMCs which shows the strengthening effect of a higher number of type-4 boundaries. Figure 8.9 compares the force-displacement curves of both fully and partially oxidised

SMC to a typical curve for Somaloy 500 [32]. The curve progression of both SMC strongly differs, particularly the maximum force of Somaloy 500 is with about 36 N almost a magnitude lower. For Somaloy 500, the SEVNB method was used. The specimens of a larger geometry as in this study (45 x 4 x 3 mm^{3}) were tested in a four-point bending setup with an inner span of 20 mm and an outer span of 40 mm at a cross head speed of 0.05 mm/min. It is reported that the crack growth started at the maximum load for Somaloy 500. In Somaloy 700 3P, however, the crack formation sometimes started at earlier stages as observed in the videos from the CCD camera.

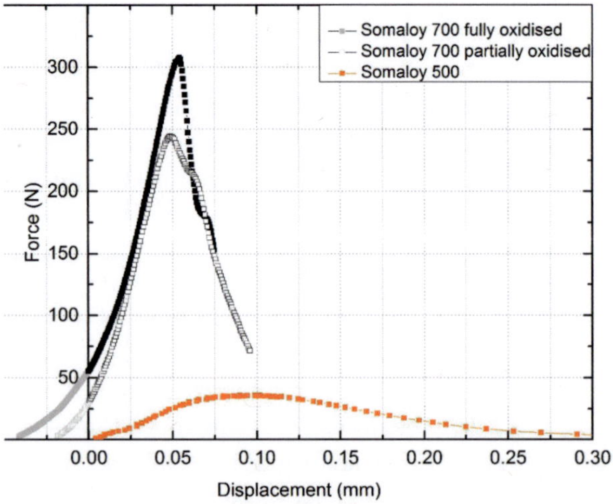

Figure 8.9: Force-displacement curve for Somaloy 500 and fully and partially oxidised Somaloy 700 3P. Fully oxidised Somaloy 700 3P reaches the highest forces whereas Somaloy 500 sustains by far the lowest force. The data for Somaloy 500 are replotted from [32].

The fracture surfaces of micro- and macromechanical tests appear similar for type-1 to -3 boundaries. Type-4 boundaries show a different fracture surface at the macroscale. For type-1 to -3 boundaries, the same smooth fracture surfaces and cracks were observed in macroscopic four-point bending as well as in microtensile and microcompression testing. Type-4 boundaries showed a ductile behaviour at the microscale and could not be tested up to failure. In macroscopic four-point bending, a bridging effect was observed only for type-4 boundaries. It is argued that the bridge is formed only at type-4 boundaries because the adhesion between the iron particle and the iron oxide layer is stronger than the adhesion between the particle and the phosphorous layer. The additional porous iron oxide layer may increase the adhesion between the particles and the other boundary layers and, thus, cause the bridging leading to the more ductile behaviour of the type-4 boundary (see Figure 8.10). Further investigations have to be done to identify the chemical composition of the bridge but it may be argued that the bridge is formed by the ductile boundary layers themselves.

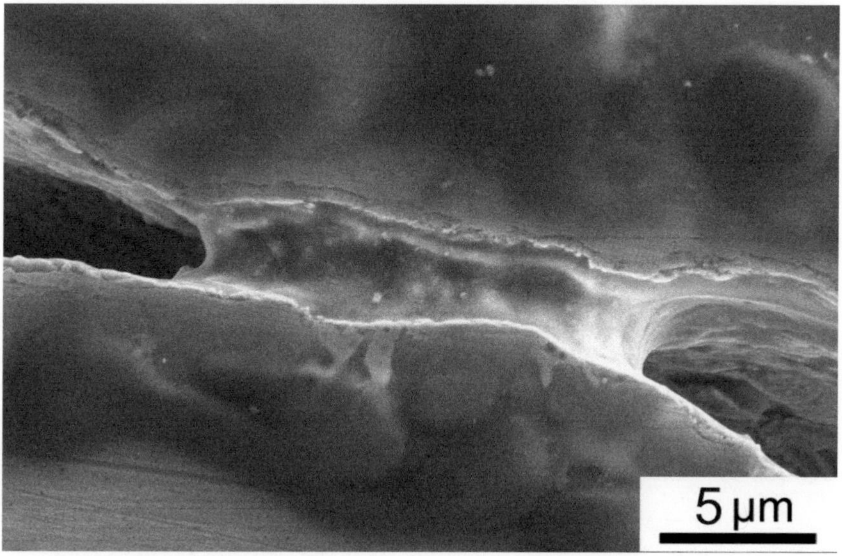

Figure 8.10: SEM micrograph of a bridge in a crack of a type-4 boundary from a fully oxidised four point bending test specimen after testing.

The strength-enhancing effect of the type-4 boundary was not only observed at the microscale but also on the macroscale.Four-point bending tests were performed in chapter 8 on fully and partially oxidised specimens. Using these results and the standard SEVNB method [88], the preliminary stress intensity factor K_Q can be calculated. The K_Q for fully oxidised SMC is 3.65 ± 0.43 MPa$\sqrt{m}$ and for partially oxidized SMC it is 3.00 ± 0.28 MPa$\sqrt{m}$. Figure 8.11 compares the K_Q to the fracture toughness of Somaloy 500.

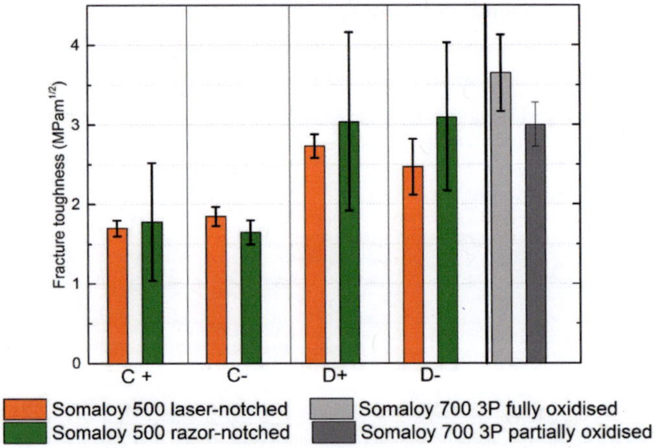

Figure 8.11: The preliminary stress intensity factor of a Somaloy 700 3P in fully and partially oxidised state is compared with the fracture toughness for Somaloy 500 from [32]. The latter was annealed at 500°C for 30 min in air (labelled with C) or at 220°C for 60 min in air (labelled with D). They were also compressed to a green density of either 7.35-7.40 g/cm3 (labelled with "+") or 7.2 g/cm3 (labelled with "-").

Four different states of Somaloy 500 were investigated in [32]: the specimens labelled with C were heat treated at 500 °C for 30 min in air whereas specimens labelled with D were heat treated at 225 °C for 60 min in air. Specimens marked with "+" had green densities of 7.35-7.40 g/cm^3 and specimens labelled with "-" had green densities of 7.20 g/cm^3. Furthermore, two different methods were used to introduce the notch: the laser cut method or the razor blade method. C specimens have the lowest fracture toughness with a maximum fracture toughness of 1.85 ± 0.12 MPa$\sqrt{m}$ using the laser cut method. The heat treatment obviously strongly influences the fracture toughness shown by the D specimens which have a maximum fracture toughness of 3.04 ± 1.15 MPa√m using the razor blade method. These values are in the range of the Somaloy 700 3P specimens. In [33], the annealing temperature for Somaloy 700 3P was increased from 530 °C to 700 °C. The structure of the boundary layer changed from amorphous to crystalline definitely changing the mechanical properties of the boundary. Thus, changing the green density, pressure or annealing temperature leads to a completely new spectrum of boundary structures with new unique properties.

The comparison of the force-displacement curves and the calculated preliminary stress intensity values of Somaloy 700 3P and the fracture toughness of Somaloy 500 [32] allows for quantifying the role of the type-4 boundary for the fracture toughness. Assuming a linear behaviour and thus a dominance of the type-4 boundary, the fracture toughness for an ideal SMC solely consisting of type-4 particle boundaries would be 4.86 MPa√m as shown in Figure 8.12. The estimate is based on only three datapoints. One data point is even taken from Somaloy 500 [32]. However, there is some theoretical support that the extrapolation is valid. In principle, one could analyse the influence of the different types of particle boundaries on the fracture toughness with the help of an adjusted rule of mixture. In [120], the fracture toughness for a metal-ceramic functionally graded material (FGM) is determined by a rule of mixture:

$$G_{IC}(x) = V_m(x) \cdot G_{IC}^{metal} + [1 - V_m(x)] \cdot G_{IC}^{ceram} \tag{8.1}$$

where $G_{IC}(x)$ is the critical energy release rate of the FGM, G_{IC}^{ceram} and G_{IC}^{metal} are the critical energy release rates of the ceramic and metal, respectively, and $V_m(x)$ is the volume fraction of the metal. This rule holds for cracks which grow through both the metal and ceramic phases without any debonding. The metal phase fractures in a brittle manner. If the metal grains are considerably larger than that of the ceramic in the FGM, a crack bridging concept can be applied leading to an R-curve behaviour. For Somaloy 700 3P, there is a transition from brittle type-1-3 boundaries to the ductile type-4 boundary and vice versa. With this approach, the dominance and importance of the type-4 boundary could be analysed. The difference between the brittle types of particle boundaries may also be considered. In addition, it could be possible to observe an R-curve behaviour for type-4 boundaries as crack bridging was observed (see Figure 8.10).

For the material design of a complex composite, i.e. Somaloy 700 3P, another useful approach is to consider the critical flaw size. The critical defect size a_c can be estimated rearranging equation (2). As the SMC exhibits a strong tension-compression asymmetry, two different yield strengths are used which were obtained from tension and bending all performed at the macroscale. K_{IC} and Y are taken from this work, whereas the tension and bending strengths are taken from literature [95].

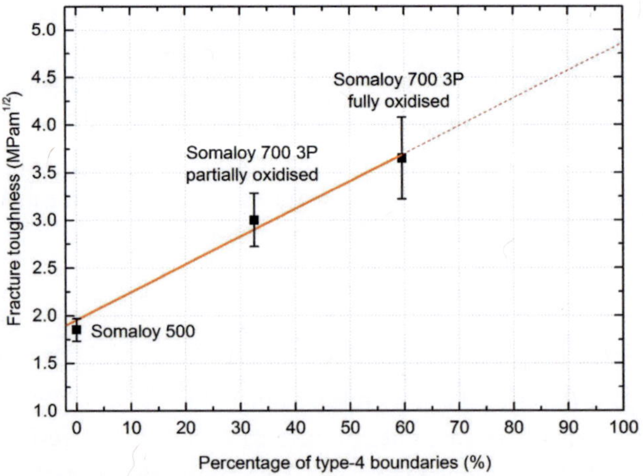

Figure 8.12: Estimation of the influence of the percentage of type-4 boundaries in Somaloy 700 3P on the fracture toughness. The fracture toughness of Somaloy 500 is taken from [32].

The critical flaw sizes in tension and bending are then:

$$a_{c_{tension}} = 1925 \mu m \tag{8.2a}$$

$$a_{c_{bending}} = 520\mu m \tag{7.2b}$$

Considering that 68% of all particles of Somaloy 700 3P are between 150-425 µm, the defect size in tension and bending is larger than the iron particles themselves. Also, the defect sizes are almost two magnitudes larger than the boundary thickness which is approximately between 20 nm and 1 µm depending on the type of boundary. This explains that on the microscale, the measured strengths are considerably higher.

The estimate of the fracture toughness of Somaloy 700 3P and its corresponding critical crack length shows that compared to Somaloy 500, the additional particle boundary leads to an improvement of the mechanical properties of the SMC. However, even the estimated fracture toughness for a hundred percent type-4 boundaries lies only in the regime of a tough ceramic. This leads to the conclusion that there is a natural limit to the improvement of the mechanical properties of the Somaloy 700 3P.

The distribution of the types of particle boundaries within the SMC plays an important role for the mechanical properties of the SMC. The point count method is a very simple and straightforward method to describe the distribution of the particle boundaries within a specimen. As shown in section 5.4, SEM images of a large enough resolution were taken along a cross section of a partially oxidised SMC. A grid of points was laid onto the images and the points that hit a boundary were categorised according to the type of boundary. As the mechanical properties of the different types of particle boundaries themselves clearly vary in bending, tension and compression on the microscale, local mechanical variations may be identified with the help of the point count method. It could also give valuable information for designing and modelling electrical machines. As an example, a full oxidation of a complex device may not be guaranteed during processing. The distribution of the types of particle boundaries can be identified by the point count method. If a region of a high content of brittle particle boundaries with reduced mechanical properties is located in a part of a component locally experiencing higher loads, device failure might occur. The device can be optimized by changing the design or adjusting the processing parameters and, thus, assuring a higher concentration of type-4 boundaries in areas of high mechanical loading.

8.1.3 The Influence of Pores on the Mechanical Behaviour of Somaloy 700 3P at the Micro- and Macroscale

Pores in general influence the mechanical properties of a solid material by usually reducing its mechanical resistance. In the SMC, the full scale of pores from a few tens of nanometres up to large, macroscopic pores were found (see Figure 8.13). The size of the macroscopic pores even reaches the millimeter range. Those pores can be partially filled with randomly distributed boundary layers. Micron-sized pores occur between particles and in the middle of a boundary layer, whereas nano-sized pores are found in the iron oxide layer of the type-4 boundary (see Figure 8.14a). Small pores also occur in the bulk particles (see Figure 8.14b and c).

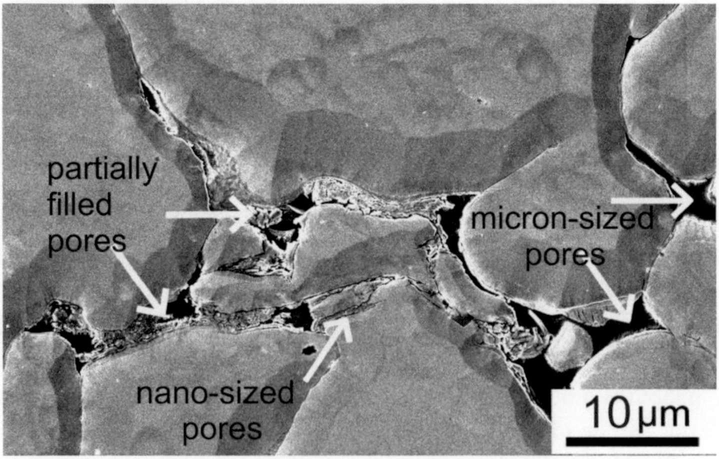

Figure 8.13: Full range of particle boundaries that are present in Somaloy 7000 3P: from macroscopic pores over micron-sized pores up to nano-sized pores in the boundary layers. The macroscopic pores and micron-sized pores may be unfilled or partially filled with boundary layers.

Three different behaviours of the pores were found: i) they contribute to the mechanical failure, ii) they do not affect the failure at all and iii) they enhance the mechanical properties of the particle boundaries.

In macroscopic four-point bending, the expected mechanical impact of the pores is observed: macroscopic pores contribute to the failure of the material (see section 7). The crack propagates along the particle boundaries and favours weak particle boundaries and the macroscopic pores because of the lower energy needed (see Figure 7.1 and Figure 7.2). In microcompression, the pillars with particle boundaries sometimes contained pores of different sizes (see section 6.4). Although the particle boundaries are not involved in the slip deformation, the presence of pores changes the overall mechanical response of the pillar. If the volume of a pore is large enough, it reduces the mechanical strength drastically up to a complete failure of the pillar (see Figure 6.6). A similar behaviour was found in the microtensile tests in section 6.6. In the microtensile test specimen with a type-2 boundary, failure did not occur at the expected boundary. Instead, the combination of a type-1 boundary with a pore was favoured in comparison to the solid type-2 boundary. One reasons why this combination was favoured is that in microbending, a type-1 boundary was shown to endure only about half the failure strength of a type-2 boundary (section 6.5).

In contradiction, pores also do not take part in a deformation mechanism at all. The small pores in the bulk particles do not influence any type of micromechanical testing (see Figure 8.14b and c). A microcantilever with a type-4 boundary and a pore did not fail (see Figure 8.14a). In general, in compression and bending at the micro-scale, the micro- and nano-sized pores were not observed to affect the mechanical behaviour.

The nanometre-sized pores in the additional iron oxide layer of a type-4 boundary forming between the particle and the phosphorous layer, contribute to the ductile behaviour of the particle boundary. In microcantilever bending, the cantilevers with mixed type-3/type-4 boundaries failed at the type-3 boundary interface whereas the type-4 boundary interface remained unaffected (see section 6.5). In microtensile testing, a bridging effect was observed only for type-4 boundaries with the assumption that the porous iron oxide layer of the type-4 boundary is responsible for the bridging effect (see Figure 7.2 and Figure 8.10). It may be argued that either the pores actively support the adhesion effect of the iron oxide layer or the layer itself causes the improved particle- boundary interface. Further investigations with micro-tensile tests or a higher resolution camera for the four-point bending tests are needed to describe the adhesive iron oxide layer of the type-4 boundary.

Although the porous iron oxide layer of the type-4 boundary is not a hierarchical porous structure, its possibly strength-enhancing effect may be compared to the findings in nature and other porous materials. Natural materials, e.g. wood or bone, typically contain hierarchical porous structures offering outstanding mechanical properties which surpass the properties of the constituent materials [121]. In [122], it is described how natural materials are insensitive to flaws at the nanoscale due to their hierarchical structure. Inspired by these natural structures, gradient porous structures are fabricated from polymers, metals, ceramics and polymers with improved thermal and mechanical properties. As an example, a graded micro- to nanoporous foam of a stiff polymer such as poly(methyl methacrylate) (PMMA) was observed to have an increasing storage modulus with increasing pore-area fraction [121]. Another example is a nanoporous open-cell gold foam. The strength of the foam is not only controlled by its relative density but also by its ligament size at the nanoscale. Grain boundaries and surfaces may give a similar strengthening behaviour [123].

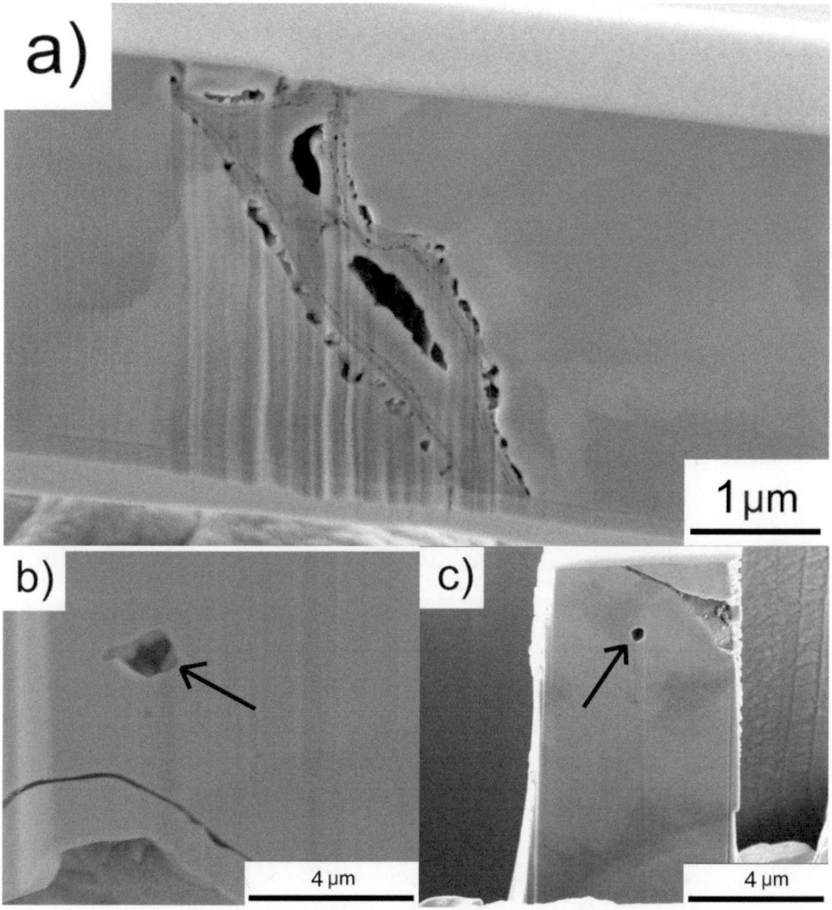

Figure 8.14: a) SEM micrograph of nano-sized pores of the iron oxide layer of the type-4 boundary and some micron-sized centre pores in the middle of the second iron-oxide layer of the boundary. b) small pore in the bulk of a uncompressed particle of Somaloy 700 3P. c) micropillar containing a small pore in the bulk iron particle.

8.1.4 Concluding Remarks

Section 8.1 addressed the mechanical properties of an SMC regarding its particles, particle boundaries and pores at different length scales. The particles seem to contain several grains with a pronounced subgrain structure due to compaction and are softer than the particle boundaries.

A model was developed to describe four different types of particle boundaries with distinctive mechanical properties at the micro-scale. It is argued that the boundary layers are hard but not very brittle and generally strengthen the SMCs. The particle-particle boundary interface critically contributes to the failure. Type-4 boundaries have a ductile particle-boundary interface due to its additional porous iron oxide layer.

Pores are found in both particles and boundaries and range from nano- to millimetres in size. They either do not take part in the deformation or failure mechanism, actively cause failure or even appear to be strength-enhancing at least to some extent at the nanoscale.

The tension-compression asymmetry at the macroscale is also found at the microscale. The iron particles determine the failure strength in compression whereas the particle boundaries determine the failure in tension.

Although the preliminary fracture toughness of Somaloy 700 3P is higher than the fracture toughness for Somaloy 500, it is generally in the regime of a tough ceramic. There is a natural limit of the mechanical properties of the whole SMC which is related to the properties of its boundaries.

8.2 Critical Evaluation of Macro- and Micromechanical Testing Methods on SMCs

Complex materials consisting of several components require the analysis of each individual component in order to understand the underlying deformation and fracture mechanisms. This applies to Somaloy 700 3P whose particles and particle boundaries were tested separately in order to explain the microscopic and macroscopic mechanical behaviour of the material. The complexity of the material makes macroscale and small scale testing difficult. The main points addressed in this section are the preparation of sample for micro- and macro-mechanical testing and the testing methods themselves.

8.2.1 Sample Preparation for Micro- and Macromechanical Testing

The preparation of mascroscale four-point bending specimens seems straight forward but requires some caution when choosing the samples and preparing the notch. The 9mm thick LTRS-bars were cut in the middle to produce the bending specimens A and B as shown in Figure 3.3. The Point Count Method showed though, that specimens A and B both cover a transition region, i.e. a gradual decrease or increase of oxidation. This leads to the very similar force-displacment diagrams in Figure 7.3. In order to analyse only the core and thus a region with almost no oxidation, the specimen needs to be cut out of the middle of the LTRS-bars. It is recommended to first analyse the oxidation gradient by the suggested Point Count Method and then to choose the specimens. The razor blade method is used to introduce a sharp notch in brittle materials and the crack is supposed to start from the notch root. In Somaloy 700 3P, the crack does not start at the sharp notch root but at the particle boundaries. Figure 8.15 shows both fracture surfaces side a and side b of the same specimen after testing. Although the razor blade notch was straight and smooth, the seemingly "rough" notch root shows that the crack was not initiated at the root but ran along the particle boundaries. The problem of notching was already stated in [113] for powder green compacts and a preparation method was suggested for Somaloy 500 in [32] which was used in this work. However, both did not polish the surface to see the crack propagation. Both polishing the specimen surfaces and investigating the fracture surfaces after testing reveal that the crack does not start at the notch root.

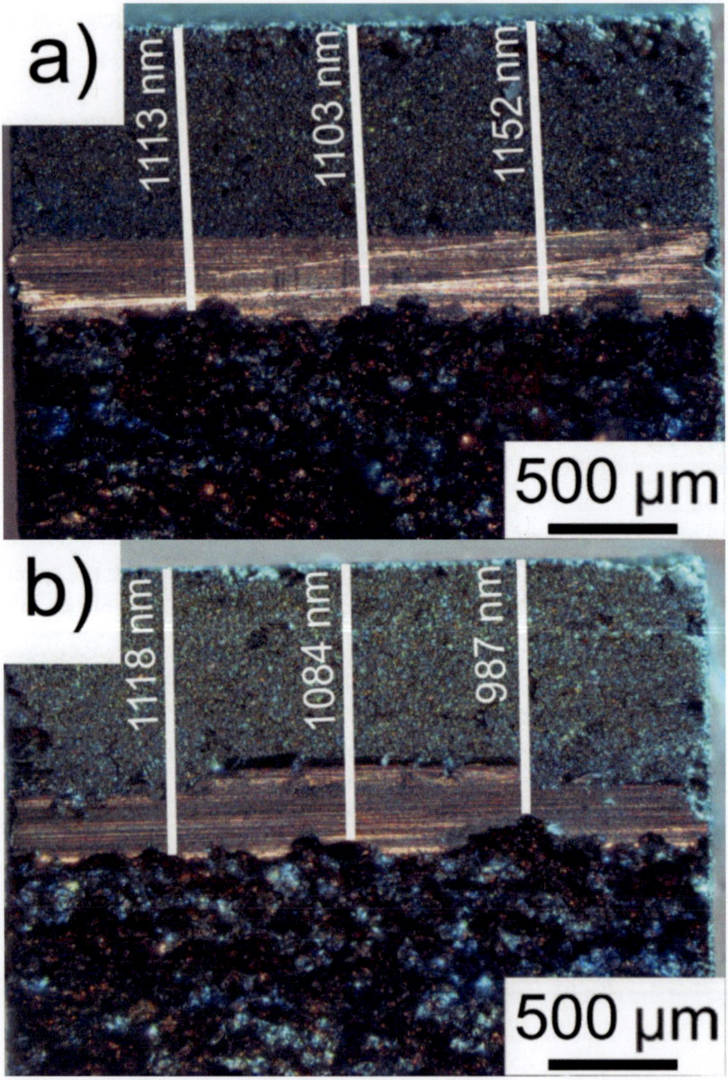

Figure 8.15: a) Fracture surface of a razor blade notched specimen. Although the razor blade notch is sharp, the crack ran along the particle boundaries, b) Other side of the same specimen. The measured crack length here is different than on the other side indicating that the crack did not run through the notch root.

The surface preparation of SMC for micromechanical testing is difficult because of the soft iron particles and the brittle boundary interfaces. A convincing method was found by combining mechanical polishing and ion beam polishing. Chapter 4 describes the preparation methods used in this work. Mechanical polishing only does not lead to a satisfactory surface quality. Although better results are obtained when using a chemo-mechanical end polishing treatment compared to a pure mechanical polishing step, the boundary layers always form a kind of relief (see section 4.1). Mechanical polishing leaves a deformation layer of about 245 nm which influences nanoindentation measurements (see section 4.3). In previous works on Somaloy 500 and 700 3P, etching, electrolytic polishing [33] or simple chemo-mechanical polishing [17] were used for preparation. All methods caused a strong surface relief. For ESBD measurement, it results in large unidentified areas (see Figure 1.8). Figure 8.16a shows a bright field image of Somaloy 700 3P etched with 3% nitric acid. The particle boundaries are attacked more by the acid causing a strong relief in the particle boundaries making the identification of individual layers difficult. A polishing with alumina-based fine-grained suspension as end polishing step (see step 7b, Table 3.3 and Figure 8.16b) does not show such a pronounced relief. The individual boundary layers can be identified. Both planar and cross-sectional ion beam polishing were shown to be successful for exposing the real microstructure and for smooth surfaces for reliable ESBD investigations but only if best-practice procedures are followed. Otherwise, it results in an over-etching at the particle boundaries and even at the particles. The recommended preparation method is a mechanical polishing up to 0.25 μm with an alcohol-based diamond suspension and an oil-water based lubricant and a ion beam polishing without a cleaning step and a voltage of 4k V for 5 h at an angle of 10.5° for planar ion beam polishing (procedure D, see section 4.2). The deformation layer observed after mechanical polishing can be completely removed by ion beam polishing.

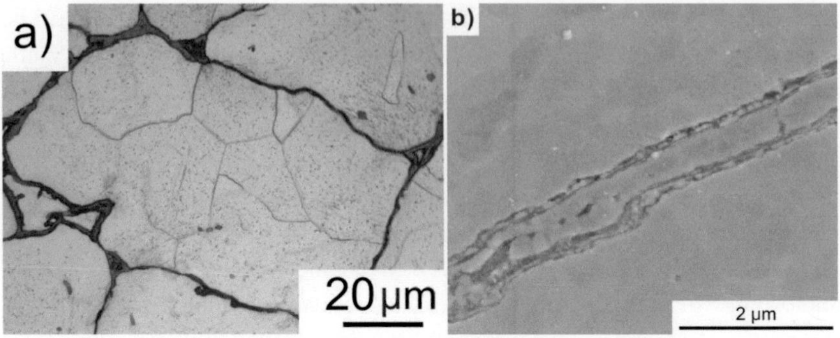

Figure 8.16: a) Bright field light microscopy image of Somaloy 700HR-3P etched with 3% nitric acid [33], b) SEM image of Somaloy 700 3P after polishing with an alumina-based fine-grained suspension.

Not only the surface preparation requires certain skills but the fabrication of the micro test structures is also difficult. All methods are time-consuming and selecting the position of the particle boundaries in the test structure is not straight-forward. On the one hand, it was easy to place a test structure because the particle boundaries are visible on the well-polished surface. On the other hand, the progression of the boundary in the volume was always unknown. In microcompression tests, most of the particle boundaries lie in the upper third of the prepared pillar and also contained pores and different types of particle boundaries (see Figure 8.6). In the microcantilevers, the position of the particle boundaries was not perpendicular to the surface and their position within the cantilever varied (see Figure 6.7). In microtensile testing, similar problems occurred considering the position of the particle boundaries within the tensile specimen, but in this case the surrounding material also influenced the tests. It is nonetheless possible to produce tests structures on complex materials such as the SMC. Test structures on materials such as nanoporous open cell foams [123] and ceramic matrix composites [124] were also successfully produced and tested. For the latter, the structure of the composite was more ordered compared to the SMC making it possible to place SiC fibres within a SiC matrix along the shear plane in a micropillar in order to test the properties of the fibre-matrix interface.

8.2.2 Macroscopic Four-Point Bending Tests

There are two restricting effects to the method used for calculating the fracture toughness of Somaloy 700 3P: i) the notch root does not fulfil the requirements of the method and ii) not all of the particles are brittle. In this work, the SEVNB method was used for preparing test specimens and calculating the fracture toughness. As explained in section 8.2.1, the introduction of a crack at a specified notch root was not possible for Somaloy 700 3P affecting the calculation of the fracture toughness. As stated by equations (10)-(12) in section 3.12, the crack must fulfil certain conditions so that the testing method is valid [89]. When determining the crack lengths according to equation (10) and (11), side b fails the criteria but side a is valid (see Table 8.1). This shows how important a precise notch root is for calculating the fracture toughness. There will be a certain error due to the poor quality of the notch root.

Table 8.1: Criteria for determining the crack length.

Criteria	side a	valid	side b	valid
$a = \frac{(a_1 + a_2 + a_3)}{3}$	1123 nm		1063 nm	
$\frac{(a_{\max} - a_{\min})}{a} \leq 0.1$	0.04	yes	0.12	no
$0.2 < \alpha = \frac{a}{W} < 0.3$	0.28	yes	0.26	yes

At first, the particle boundaries were assumed to be brittle. Because of that, an approach for brittle i.e. ceramic material was chosen to determine the fracture toughness. The type-4 boundaries, though, show a ductile behaviour changing the approach to calculate a fracture toughness. For applying LEFM, a purely brittle behaviour of the material is required. Although the type-1 to 3 boundaries are brittle, the type-4 boundary behaves in a ductile manner and its impact on the fracture toughness is not fully investigated yet as stated in section 8.1.2. Even crack bridging was observed in the macroscopic four-point bending tests which would suggest an R-curve behaviour of the type-4 boundaries. It is nonetheless possible to give a rough estimation of the fracture toughness of Somaloy 700 3P in terms of a preliminary stress intensity factor K_Q calculated by the SEVNB method.

8.2.3 Nanoindentation

Nanoindentation on a complex material such as SMC can be performed if certain aspects are considered: the influence of a possible deformation layer from sample preparation and the influence of the subsurface. Ion beam polishing is required to fully remove the deformation layer from mechanical polishing. The best-practice polishing method (procedure D) leads to a smaller indentation size effect than polishing methods that did not remove the deformation layers. At lower indentation depths, the effect is even more pronounced. The subsurface of the SMC strongly influences the nanoindentation measurements. In section 6.2, the indentation along a boundary showed that both hardness and Young's modulus are significantly lower for three specific indents. The cross section through one of these indents reveals a large pore placed directly underneath the indent (see Figure 8.17). This explains the drops in both hardness and Young's modulus. Thus, the measured values are not boundary-specific but highly influenced by the subsurface. *In general, such indentation results must not be included in an evaluation without a thorough local investigation of the material tested. It is difficult to resolve boundary-specific properties with nanoindentation.* [41]

Figure 8.17: Indent 9 clearly hit the boundary and yielded a much reduced hardness and modulus. Cross-sectional analysis of indent 9 revealed a large pore underneath the indent which very likely caused the decrease in mechanical properties [41]

The influence of the subsurface is also observed when measuring bulk particles by nanoindentation. Figure 8.18a replots the data from Figure 6.2 in form of a hardness map for an indentation depth of 300 nm. While the deformation layer was removed by the procedures provided in this work, the influence of the subsurface cannot be removed. Figure 8.18b shows the stiffness squared over force over displacement diagram. This ratio can be used to find the machine compliance and to calibrate the area function when using the CSM method [125]. Plotting the parameter over the displacement should result into a flat curve, as the ratio only depends on the reduced modulus and hardness. Thus, the ratio provides a method to reassure if the machine compliance is correctly analysed. All curves that do not reach a plateau value must be tagged because the subsurface influenced the measurement. Removing all the affected measurements, only 42 indentations are left that showed no influence of the subsurface (see Figure 8.18c). Although the CSM method can be used to determine the hardness of the bulk particles, it is not possible to resolve subgrain structures formed during compaction. In [17], nanoindentation was used to identify the recovery progress of the compressed bulk iron particles of an SMC of the Somaloy family for different annealing temperatures. The samples were chemo-mechanical polished and indented with an indentation depth of 200 nm with an unspecified method. From the experiences collected in this work on indenting a SMC, there are two remarks: i) the used preparation method and the indentation depth of 200 nm leads to higher hardness measurements and ii) the measurements vary if the quasi-static method or CSM method is used. In [17], the method was not specified, but in this work the CSM method was used. In [126], the CSM method is critically reviewed. While measuring the hardness and Young's modulus of a material by the CSM method, there can be a significant error for material with a high contact stiffness. The hardness is particularly sensitive to this problem. In [127], a significant hardness decrease is reported especially for indentation depths below 200 nm when using the CSM method compared to the quasi-static measurement. So indentation measurements below 200 nm are difficult and it is thus not possible to identify a subgrain structure within the deformed iron particles resulting from deformation processes.

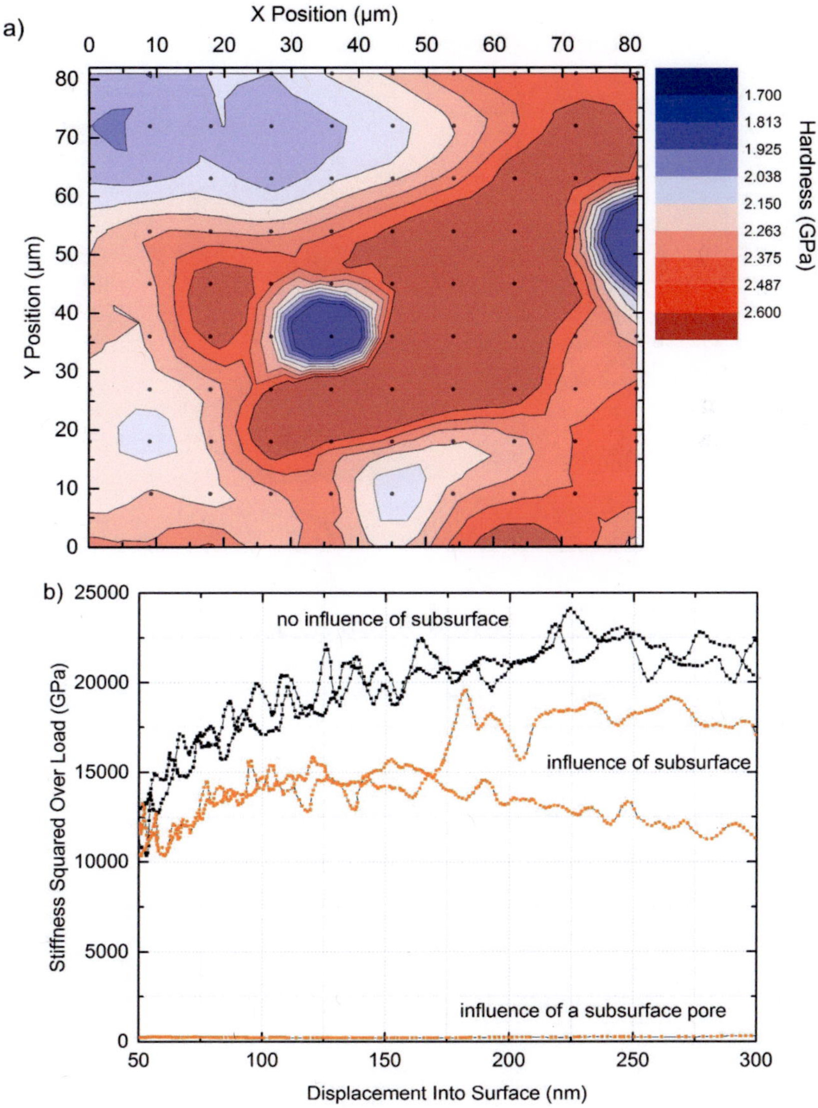

(Figure 8.18 continued on next page)

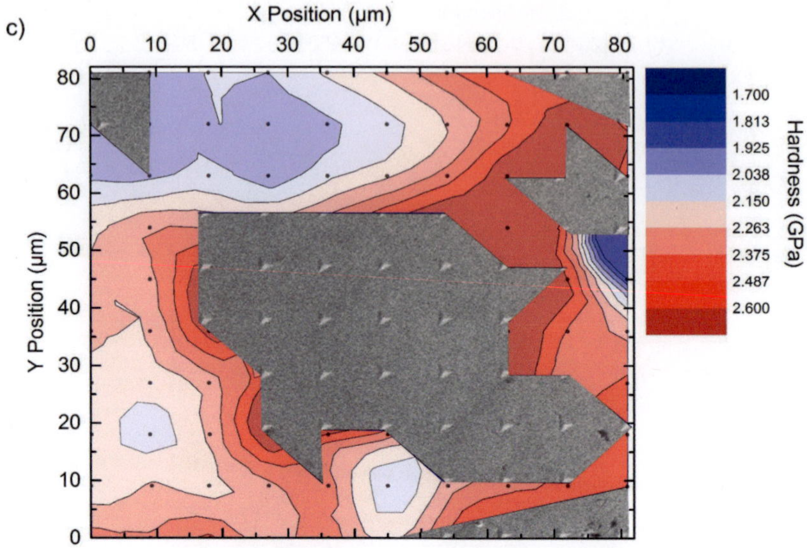

Figure 8.18: a) Hardness map for a 10 by 10 indent field at an indentation depth of 300 nm in an iron particle of the SMC with the CSM method with the influence of the subsurface. b) Stiffness squared over load vs. displacement into surface. All measurements that do not reach a plateau value must be tagged. c) Only 42 indents were left after removing the measurements influenced by the subsurface.

8.2.4 Microcompression Testing and Microcantilever Bending Testing

With micromechanical testing, a large amount of qualitative information on the mechanical behaviour of SMCs can be collected; however, it is difficult to obtain quantitative data for such complex materials. The quantitative analysis is restricted e.g. by the complexity of the material, by limited sample geometries, by the amount of specimens needed, by the amount of material required and by efficiency considering time and costs.

When performing small-scale testing, there are also some challenges with the test methods themselves. In microcompression, the effect of the taper and the subsurface are neglectable for measurements on Somaloy 700 3P. The pillar geometry in terms of taper and aspect ratio, the influence of the substrate, the possible influence of a gallium contamination of the specimen due to FIB fabrication and the misalignment or friction between indenter and pillar are of concern (see section 3.10). Equation (3) compensates the effect of the substrate and the taper of the pillar. Figure 8.19 shows an engineering stress – engineering strain curve for a pillar prepared in the bulk. The red curve represents the corrected measurement according to equation (3). For the measurements in this work, the correction is not even needed as the deviation of corrected and uncorrected curve lies in the scatter of the tested micropillars. This also supports the good quality of the performed experiments and makes evaluation straightforward and simple.

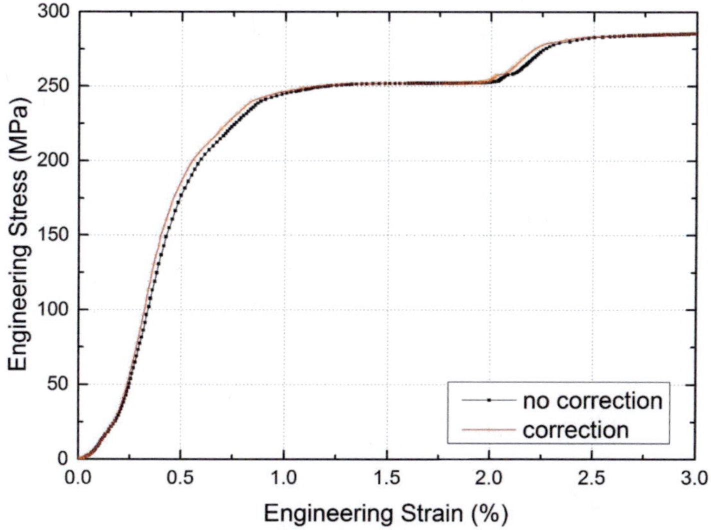

Figure 8.19: Engineering stress-engineering strain curve for a pillar in the bulk iron particles of Somaloy 700 3P with and without the correction of several constraints.

In microcantilever bending, there are more constraints. The aspect ratio of the fabricated microcantilevers from Somaloy 700 3P is below 6 meaning that the simple beam theory cannot be applied and a quantitative analysis of Young's modulus is not possible. The effect of the indentation mark on the microcantilever can be neglected. It is nonetheless possible to obtain qualitative information about the mechanical behaviour of the bulk particles and the different types of particle boundaries. In order to apply simple beam theory on microcantilevers, the loaded beam length L must be increased so that the effect of the slope becomes less important. This condition is fulfilled if the aspect ratio of the cantilevers is larger than 6. Methods for estimating the curvature effect about a point loading on a cantilever beam were given in [68, 109, 110]. In [68], a method was described to analyse the anisotropy in Young's modulus of copper by microcantilever bending (see also section 3.10) which was also used in this work and others. Triangular shaped microcantilevers with an aspect ratio larger than 6 are fabricated from copper. The cantilevers are either single loaded or multiple loaded. The beam compliance is approximated by

$$C = \frac{1}{E \cdot I_y} \cdot L^3 + m_0 \cdot L^2 \tag{8.3}$$

where C is the Compliance, E Young's modulus, I_y the moment of inertia, L the length from beginning of the cantilever to the loading point and m_0 the slope at the fixed end due to unit bending moment. The loaded beam length L must be increased so that the effect of the slope, i.e. the m_0-term, becomes less important and the simple beam theory can be applied. To test this behaviour on pure iron particles of Somaloy 700 3P, microcantilevers were fabricated in the bulk particles. The aspect ratios of the prepared were between 1 and 6. Larger aspect ratios and more experiments would be necessary to prove a dominance of the L^3-term and thus fulfil simple beam theory which assumes long, slender beams.

The influence of the imprint of the wedge tip may affect the bending measurement, but on Somaloy 700 3P the effect is neglectable. The maximum measured load on single-loaded microcantilevers is about 0.4 mN. For testing the influence of the wedge tip, it was indented in a bulk iron particle and the load-displacement curve for this experiment is given in Figure 8.20. For a maximum load of 0.4 mN, the wedge indents the cantilever about 60 nm. The maximum compliance of a single-loaded microcantilever was 0.0028 m/N which correspond to a force of 0.071 mN when deflected by 2000 nm. There is thus no influence of the wedge on the measurement.

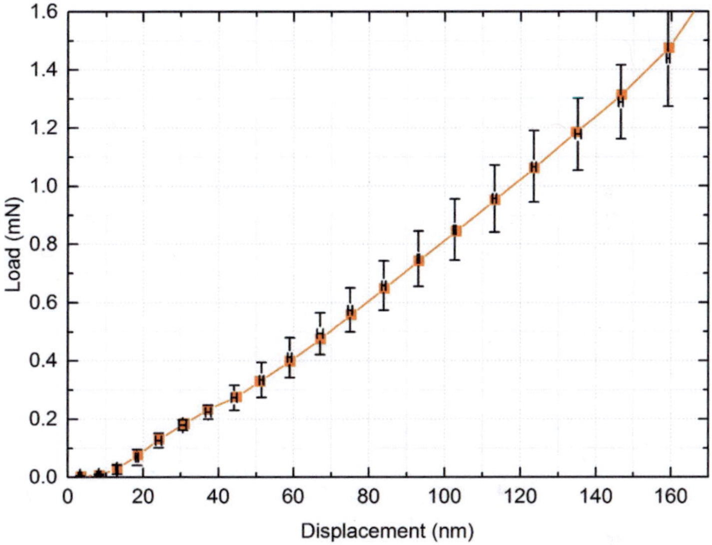

Figure 8.20: Load versus displacement curve of the wedge tip indented into a bulk iron particle of Somaloy 700 3P.

8.2.5 Concluding Remarks

This section addressed the challenges of the micro- and macromechanical testing of the complex composite powder metal Somaloy 700 3P. All methods required a careful sample handling considering cutting, placement and metallographic preparation. The examined surfaces must be at least mechanically polished for macromechanical testing and ion beam polished for micromechanical testing in order to reveal the microstructure in terms of particles, pores and particle boundaries.

Macroscopic four-point bending tests on Somaloy 700 3P required a careful preparation of the notch for which no suitable method is found yet. The SEVNB method may not be applicable to Somaloy 700 3P because the type-4 boundaries behave ductile and thus may show an R-curve behaviour. Nonetheless, a preliminary fracture toughness could be estimated. Furthermore, the natural limitation of the mechanical properties of Somaloy 700 3P is given by the properties and number of the type-4 boundaries.

In nanoindentation, the subsurface and a possible deformation layer from polishing influence the measurements. These influences can be easily eliminated by a proper sample preparation and a careful data evaluation. Quantitative data for Young's modulus and hardness can, thus, be determined. For microcompression, the testing method is simple to implement and also delivers quantitative results. For microcantilever testing and microtensile testing, the methods are reliable and a qualitative evaluation is possible.

The findings on the microscale and the knowledge of the mechanical properties at the macroscale can be combined though only a qualitative understanding of the size effects particularly in deformation and fracture mechanism can be obtained.

9 Summary

Soft Magnetic Composites (SMC) are important materials for the fabrication of electric machines. They consist of pure iron particles which are coated with an organic or inorganic layer. The particles are compressed and then heat treated. In this research work, the commercially available Somaloy 700 3P was investigated because it does not only have the unique magnetic properties of an SMC but also shows improved mechanical properties compared to other magnetic powder metals. During the heat treatment in water-vapour, Somaloy 700 3P forms an additional iron oxide layer which is responsible for the enhanced mechanical properties. The analysis of these boundaries requires micro-mechanical testing.

The preparation of complex composites such as the Somaloy 700 3P is quite difficult as this SMC consists of soft iron particles and brittle boundaries. A preparation method for micro-mechanical testing was developed in this research work. It consists of mechanical polishing and a consecutive ion beam polishing to reveal the real microstructure of the SMC.

The material was characterised by light microscopy, SEM, TEM, EDX and EBSD. With the combination of all of these methods, a model was developed to describe four different types of particle boundaries with distinctive mechanical properties at the micro-scale. The particle boundaries consist of several different layers containing phosphorous, oxygen and iron and show an increasing complexity depending on the degree of oxidation. The particles contain several grains with a pronounced subgrain structure. Pores are represented in the material from the nano- up to the millimetre scale.

The micromechanical properties of Somaloy 700 3P were investigated by nanoindentation, microcompression, microcantilever bending and microtensile testing. The particles are softer than the particle boundaries. It is argued that the boundary layers are hard but not very brittle and generally strengthen the SMCs. The particle-particle boundary interface is critical in terms of failure. Pores do not take part in the deformation or failure mechanism, actively cause failure or even appear to be strength-enhancing at least to some extent at the nanoscale. The tension-compression asymmetry described at the macroscale is also present at the microscale. The particles determine the failure in compression whereas in tension, the particle boundaries determine the failure strength. The preliminary fracture toughness of Somaloy 700 3P is higher than the fracture toughness for Somaloy 500. However, even with a 100 percent of type-4 boundaries, the fracture toughness of Somaloy 700 3P would be only in the regime of a tough ceramic giving a natural limit of the improvement of the mechanical properties. Besides the material characterisation and the analysis of the mechanical properties of Somaloy 700 3P, the sample preparation was also part of this work. All methods required a careful sample handling considering cutting, placement and metallographic preparation. The examined surfaces must be at least mechanically polished for macromechanical testing and ion beam polished for micromechanical testing in order to reveal the microstructure in terms of particles, pores and particle boundaries.

The SEVNB method used in macroscopic four-Point bending testing may not be applicable to Somaloy 700 3P because its type-4 boundaries show a ductile behaviour. Nonetheless, a preliminary fracture toughness could be estimated.

In nanoindentation, the subsurface and a possible deformation layer from polishing influence the measurements. These influences can be eliminated by a proper sample preparation and a careful data evaluation. Young's modulus and hardness were determined for the particles. The particle boundaries are not resolvable in nanoindentation.

For microcompression, the testing method is simple to implement and also delivers quantitative results. The influence of the substrate and the taper of the pillar were neglectable. For microcantilever testing, only qualitative results were obtained. The imprint of the wedge did not influence the measurements.

In conclusion, the methods are all applicable to Somaloy 700 3P. With a careful sample preparation, quantitative results can be obtained. Furthermore, the combination of micro and macroscale results is also possible, though it only leads to a qualitative understanding of the size effects, particularly for deformation and fracture.

Acknowledgement

This doctoral thesis originated from my work at the Institute for Applied Materials (IAM-WBM) at the Karlsruhe Institute of Technology from June 2013 to September 2017. I would like to thank everyone who has supported me in successfully completing my thesis.

A special thank you to Prof. Dr. Oliver Kraft who accepted me as his PhD student and who is also the first examiner of my work. He has supported me in all past and future projects and has given me the opportunity to get in touch with the international Materials Sciences community.

Furthermore, I would like to thank Dr. Ruth Schwaiger who has not only supported all the experimental work but has also shared her immense expertise. Moreover, she has guided me to become a better researcher.

I would like to thank all my colleagues from IAM-WBM with a special thank you to Daniela Exner, Ewald Ernst and Dimitri Litvinov for their technical and Jana Herzog for her organizational support. I would also like to thank Thomas Kreuter, Stefan Weyand, and Moritz Wenk for their experimental support.

A special thank you to Prof. Dr. Michael J. Hoffmann who is not only the second reviewer of this work but also let me work in his institute as a guest. I would like to thank all the members of IAM-KWT, in particular Dr. Susanne Wagner and Dr. Theo Fett for vivid discussions. I would like to thank Dr. Marcus Müller and Pascal Hettich for their experimental support. For their technical support, I would like to thank Dominic Creek and Rainer Müller.

I would like to thank Prof. Kazuki Takashima and Dr. Yoji Mine from Kumamoto University for their experimental support and the possibility to visit the setup on-site.

This work was supported by the Robert Bosch Foundation. From Robert Bosch GmbH, I would like to thank Dr. Witold Pieper, Dr. Jens Burghaus, Dr Frank

Felten, Dr. Arne Huber and Huong Giang Nguyen for providing samples and supporting me with inspiring discussions.

I would like to thank Meera Mani, Renata Palyafari and Susanne Hartwig for their inspirational dance classes that have kept me not only physically fit but also mentally on track.

I would like to thank Andrea Kreisel for proofreading my work.

Finally, I would like to thank my mother and my grandparents for always supporting me in whatever I undertake. Thank you to my cat Yin Yin for her unconventional help with my work.

Karlsruhe, July 2017 Tabea Gisela Schwark

Bibliography

1. Giménez, S., T. Lauwagie, G. Roebben, W. Heylen, and O. Van der Biest, *Effects of microstructural heterogeneity on the mechanical properties of pressed soft magnetic composite bodies.* Journal of Alloys and Compounds, 2006. **419**(1–2): p. 299-305.

2. Schlieper, G., *2013 Hagen Symposium: Challenges in the manufacture of soft magnetic PM components.* POWDER METALLURGY REVIEW 2014. **3**(1).

3. Shokrollahi, H. and K. Janghorban, *Soft magnetic composite materials (SMCs).* Journal of Materials Processing Technology, 2007. **189**(1–3): p. 1-12.

4. V. Iancu, T.C., D.C. Popa, L. Szabó. *Soft Magnetic Composites Used for the Iron Core of the Electrical Machines.* in *Proceedings of the 4th International Conference on Materials and Manufacturing Technologies*. 2006. Cluj-Napoca, Romania.

5. Fish, G.E., *Soft magnetic materials.* Proceedings of the IEEE, 1990. **78**(6): p. 947-972.

6. Lefebvre, L.P., S. Pelletier, and C. Gélinas, *Effect of electrical resistivity on core losses in soft magnetic iron powder materials.* Journal of Magnetism and Magnetic Materials, 1997. **176**(2): p. L93-L96.

7. Lambeck, M., *Barkhausen-Effekt und Nachwirkung in Ferromagnetika sowie analoge Erscheinungen in der Festkörperphysik* 1971, Berlin: Walter de Gruyter & CO. 149.

8. Liu, X., Y. Wang, J. Zhu, Y. Guo, G. Lei, and C. Liu, *Calculation of core loss and copper loss in amorphous/nanocrystalline core-based high-frequency transformer.* AIP Advances, 2016. **6**(5): p. 055927.

9. Krings, A., A. Boglietti, A. Cavagnino, and S. Sprague, *Soft Magnetic Material Status and Trends in Electric Machines*. IEEE Transactions on Industrial Electronics, 2017. **64**(3): p. 2405-2414.

10. Lars-Olov Pennander, A.G.J. *Soft Magnetic Iron Powder Materials AC Properties and their Application in Electrical Machines*. in *EURO PM2003*. 2003. Valencia.

11. Nowosielski, R., *Soft magnetic polymer-metal composites consisting of nanostructural Fe-basic powders*. Journal of of Achievements in Materials and Manufacturing Engineering, 2007. **24**(1): p. 68-77.

12. R. Bureš, M.F., P. Kollár, J. Füzer, M. Strečková, *Microstructure and fracture of magnetic composites with vitroperm addition*. Powder Metallurgy Progress, 2012. **12**(3): p. 181-186.

13. Magdaléna Strečková, M.F., Radovan Bureš, Eva Dudrová. *Improvement of mechanical properties of SMC through different hybrid inorganic-organic insulating coatings*. in *Acta Metallurgica Slovaca - Conference*. 2013.

14. P. Gramatyka, R.N., P. Sakiewicz, T. Raszka, *Soft magnetic composite based on nanocrystalline Fe73.5Cu1Nb3Si13.5B9 and Fe powders*. Journal of Achievements in Materials and Manufacturing Engineering, 2006. **15**(1-2): p. 27-31.

15. Schoppa, A., P. Delarbre, E. Holzmann, and M. Sigl. *Magnetic properties of soft magnetic powder composites at higher frequencies in comparison with electrical steels*. in *2013 3rd International Electric Drives Production Conference (EDPC)*. 2013.

16. Hultman, L.O. and A.G. Jack. *Soft magnetic composites-materials and applications*. in *Electric Machines and Drives Conference, 2003. IEMDC'03. IEEE International*. 2003.

17. Oikonomou, C., D.C. Gutiérrez, M. Monclús, J.M. Molina-Adareguia, and L. Nyborg, *Assessment of the compacting and annealing process steps on the performance of finished Soft Magnetic Composite components*, in *T2 Euro PM2015 Proceedings*2015.

18. Taghvaei, A.H., H. Shokrollahi, and K. Janghorban, *Properties of iron-based soft magnetic composite with iron phosphate–silane insulation coating*. Journal of Alloys and Compounds, 2009. **481**(1–2): p. 681-686.

19. Yiping, D., G. Youguang, and Z. Jianguo. *Investigation of motor topologies for SMC application*. in *Electrical Machines and Systems, 2007. ICEMS. International Conference on*. 2007.

20. Svensson, L., K. Frogner, P. Jeppsson, T. Cedell, and M. Andersson, *Soft magnetic moldable composites: Properties and applications*. Journal of Magnetism and Magnetic Materials, 2012. **324**(18): p. 2717-2722.

21. Henaux, C., B. Nogarede, and D. Harribey, *A New Concept of Modular Permanent Magnet and Soft Magnetic Compound Motor Dedicated to Widespread Application*. IEEE Transactions on Magnetics, 2012. **48**(6): p. 2035-2043.

22. Schwark, T., *Ashby Diagramm*, in *CES EduPack 2013 – Granta Design Limited*2015.

23. Ishikawa, T., K. Takahashi, Q.V. Ho, M. Matsunami, and N. Kurita, *Analysis of Novel Brushless DC Motors Made of Soft Magnetic Composite Core*. IEEE Transactions on Magnetics, 2012. **48**(2): p. 971-974.

24. Huang, Y., J. Zhu, Y. Guo, and Q. Hu. *Development of a High-Speed Claw Pole Motor with Soft Magnetic Composite Core*. in *2007 IEEE International Electric Machines & Drives Conference*. 2007.

25. Kim, C.W., G.H. Jang, J.M. Kim, J.H. Ahn, C.H. Baek, and J.Y. Choi, *Comparison of Axial Flux Permanent Magnet Synchronous Machines with Electrical Steel Core and Soft Magnetic Composites Core*. IEEE Transactions on Magnetics, 2017. **PP**(99): p. 1-1.

26. O. Andersson, P.H. *Advances in Soft Magnetic Composites – Materials and Applications*. in *PowderMet2009*. 2009 Las Vegas.

27. M. Persson, G.N., and A. G. Jack *Soft magnetic composite materials ac properties and their application in electrical machines*, in *Proc. Int. Conf. Electr. Mach. Syst.*2004: Cheju-Island, Korea.

28. Oikonomou, C., E. Hryha, and L. Nyborg, *Development of methodology for surface analysis of soft magnetic composite powders.* Surface and Interface Analysis, 2012. **44**(8): p. 1166-1170.

29. Shin Tajima, T.H., Mikio Kondoh, Masaki Sugiyama, Kiyoshi Higashiyama, Hidefumi Kishimoto and Tadayoshi Kikko, *Properties of High Density Magnetic Composite (HDMC) by Warm Compaction Using Die Wall Lubrication* Materials Transactions, 2004. **45**(6): p. 1891-1894.

30. Szabó, L., I. Viorel, V. Iancu, and D. Popa, *Soft magnetic composites used in transverse flux machines.* Oradea University Annals, Electrotechnical Fascicle, 2004: p. 134-141.

31. Gilbert, I., S. Bull, T. Evans, A. Jack, D. Stephenson, and A. De Sa, *Effects of processing upon the properties of soft magnetic composites for low loss applications.* Journal of Materials Science, 2004. **39**(2): p. 457-461.

32. Giménez, S., S. Herter, R. Kott, and O. Van der Biest. *Determination of the fracture toughness of soft magnetic composites by the single-edge-V-notched-beam method.* in *European Congress and Exhibition on Powder Metallurgy*. 2005 Prague, Czech Republic: EPMA.

33. Levin, M.C., *Charakterisierung von weichmagnetischen Pulververbundwerkstoffen für den Einsatz in Traktionsantrieben*2014: Verlag Dr. Hut.

34. Marcin Karbowiak, B.J., Dariusz Kapelski, Marek Przybylski, Barbara Slusarek. *Influence of grain size on mechanical properties of soft magnetic composites.* in *Euro PM2012 – PM Functional Materials*. 2012.

35. H. G. Nguyen, G.D., A. Hartmaier, *Grenze der Einsetzbarkeit eines weichmagnetischen Pulververbundwerkstoffes aus Sicht der Mechanik*, in *19. Symposium Verbundwerkstoffe und Werkstoffverbunde* 2013: Karlsruhe.

36. Pharr, G.M., E.G. Herbert, and Y. Gao, *The Indentation Size Effect: A Critical Examination of Experimental Observations and Mechanistic Interpretations.* Annual Review of Materials Research, 2010. **40**(1): p. 271-292.

37. Nix, W.D., *Mechanical properties of thin films.* Metallurgical Transactions A, 1989. **20**(11): p. 2217.

38. Oliver, W.C. and G.M. Pharr, *An improved technique for determining hardness and elastic modulus using load and displacement sensing indentation experiments.* Journal of Materials Research, 1992. **7**(06): p. 1564-1583.

39. Pathak, S., J. Michler, K. Wasmer, and S. Kalidindi, *Studying grain boundary regions in polycrystalline materials using spherical nano-indentation and orientation imaging microscopy.* Journal of Materials Science, 2012. **47**(2): p. 815-823.

40. Lian, J., J.E. Garay, and J. Wang, *Grain size and grain boundary effects on the mechanical behavior of fully stabilized zirconia investigated by nanoindentation.* Scripta Materialia, 2007. **56**(12): p. 1095-1098.

41. Schwark, T., M. Müller, Y. Mine, T. Kreuter, O. Kraft, and R. Schwaiger, *Preparing Soft Magnetic Composites for Structural and Micromechanical Investigations.* Practical Metallography, 2017. **54**(6): p. 366-387.

42. Schuh, C.A., *Nanoindentation studies of materials.* Materials Today, 2006. **9**(5): p. 32-40.

43. Fischer-Cripps, A.C., *Nanoindentation*2013: Springer New York.

44. Nili, H., K. Kalantar-zadeh, M. Bhaskaran, and S. Sriram, *In situ nanoindentation: Probing nanoscale multifunctionality.* Progress in Materials Science, 2013. **58**(1): p. 1-29.

45. Warren, O.L., Z. Shan, S.A.S. Asif, E.A. Stach, J.W. Morris Jr, and A.M. Minor, *In situ nanoindentation in the TEM.* Materials Today, 2007. **10**(4): p. 59-60.

46. Leitner, A., V. Maier-Kiener, and D. Kiener, *Dynamic nanoindentation testing: is there an influence on a material's hardness?* Materials Research Letters, 2017: p. 1-8.

47. Pharr, G.M., W.C. Oliver, and F.R. Brotzen, *On the generality of the relationship among contact stiffness, contact area, and elastic modulus during indentation.* Journal of Materials Research, 1992. **7**(03): p. 613-617.

48. Wheeler, J.M. and J. Michler, *Invited Article: Indenter materials for high temperature nanoindentation.* Review of Scientific Instruments, 2013. **84**(10): p. 101301.

49. Li, X. and B. Bhushan, *A review of nanoindentation continuous stiffness measurement technique and its applications.* Materials Characterization, 2002. **48**(1): p. 11-36.

50. Stephan, W., B. Hannes, S. Vitalij, M. Rolf, and S. Ruth, *Structure-property-glass transition relationships in non-isocyanate polyurethanes investigated by dynamic nanoindentation.* Materials Research Express, 2016. **3**(7): p. 075019.

51. Fischer-Cripps, A.C., *A simple phenomenological approach to nanoindentation creep.* Materials Science and Engineering: A, 2004. **385**(1): p. 74-82.

52. Sebastiani, M., K.E. Johanns, E.G. Herbert, and G.M. Pharr, *Measurement of fracture toughness by nanoindentation methods: Recent advances and future challenges.* Current Opinion in Solid State and Materials Science, 2015. **19**(6): p. 324-333.

53. Luo, Z.-P., G.-P. Zhang, and R. Schwaiger, *Microstructural vortex formation during cyclic sliding of Cu/Au multilayers.* Scripta Materialia, 2015. **107**: p. 67-70.

54. Maier, V., B. Merle, M. Göken, and K. Durst, *An improved long-term nanoindentation creep testing approach for studying the local deformation processes in nanocrystalline metals at room and elevated temperatures.* Journal of Materials Research, 2013. **28**(9): p. 1177-1188.

55. FM Halliday, D.E.A., John D Murphy, Steve G Roberts, *Nanoindentation and micromechanical testing of iron-chromium alloys implanted with iron ions.* Advanced Materials Research, 2009. **59**: p. 304-307.

56. S., H., K. A., and Z. Ph., *A combined atomic force microscopy and nanoindentation technique to investigate the elastic properties of bone structural units.* European Cells and Materials (ECM), 2001.

57. Wheeler, J.M., D.E.J. Armstrong, W. Heinz, and R. Schwaiger, *High temperature nanoindentation: The state of the art and future challenges.* Current Opinion in Solid State and Materials Science, 2015. **19**(6): p. 354-366.

58. Uchic, M.D. and D.M. Dimiduk, *A methodology to investigate size scale effects in crystalline plasticity using uniaxial compression testing.* Materials Science and Engineering: A, 2005. **400–401**: p. 268-278.

59. Kiener, D., C. Motz, and G. Dehm, *Micro-compression testing: A critical discussion of experimental constraints.* Materials Science and Engineering: A, 2009. **505**(1–2): p. 79-87.

60. Lehrer, C., L. Frey, S. Petersen, M. Mizutani, M. Takai, and H. Ryssel. *Defects and gallium-contamination during focused ion beam micro machining.* in *2000 International Conference on Ion Implantation Technology Proceedings. Ion Implantation Technology – 2000 (Cat. No.00EX432)*. 2000.

61. Uchic, M.D. and D.M. Dimiduk, *A methodology to investigate size scale effects in crystalline plasticity using uniaxial compression testing.* Materials Science and Engineering: A, 2005. **400**: p. 268-278.

62. Fei, H., A. Abraham, N. Chawla, and H. Jiang, *Evaluation of Micro-Pillar Compression Tests for Accurate Determination of Elastic-Plastic Constitutive Relations.* Journal of Applied Mechanics, 2012. **79**(6): p. 061011-061011.

63. Uchic, M.D., P.A. Shade, and D.M. Dimiduk, *Plasticity of Micrometer-Scale Single Crystals in Compression.* Annual Review of Materials Research, 2009. **39**(1): p. 361-386.

64. Di Maio, D. and S.G. Roberts, *Measuring fracture toughness of coatings using focused-ion-beam-machined microbeams.* Journal of Materials Research, 2005. **20**(02): p. 299-302.

65. Bechtle, S., H. Özcoban, E.D. Yilmaz, T. Fett, G. Rizzi, E.T. Lilleodden, N. Huber, A. Schreyer, M.V. Swain, and G.A. Schneider, *A method to determine site-specific, anisotropic fracture toughness in biological materials.* Scripta Materialia, 2012. **66**(8): p. 515-518.

66. Iqbal, F., J. Ast, M. Göken, and K. Durst, *In situ micro-cantilever tests to study fracture properties of NiAl single crystals.* Acta Materialia, 2012. **60**(3): p. 1193-1200.

67. Motz, C., T. Schöberl, and R. Pippan, *Mechanical properties of micro-sized copper bending beams machined by the focused ion beam technique.* Acta Materialia, 2005. **53**(15): p. 4269-4279.

68. Armstrong, D.E.J.W., Angus J.; Roberts, Steve G, *Measuring anisotropy in Young's modulus of copper using microcantilever testing.* Journal of Materials Research, 2009. **24**(11): p. 2.

69. Armstrong, D.E.J., A.J. Wilkinson, and S.G. Roberts, *Micro-mechanical measurements of fracture toughness of bismuth embrittled copper grain boundaries.* Philosophical Magazine Letters, 2011. **91**(6): p. 394-400.

70. Tsuchiya, T., M. Hirata, N. Chiba, R. Udo, Y. Yoshitomi, T. Ando, K. Sato, K. Takashima, Y. Higo, Y. Saotome, H. Ogawa, and K. Ozaki, *Cross comparison of thin-film tensile-testing methods examined using single-crystal silicon, polysilicon, nickel, and titanium films.* Journal of Microelectromechanical Systems, 2005. **14**(5): p. 1178-1186.

71. Hemker, K.J. and W.N. Sharpe, *Microscale Characterization of Mechanical Properties.* Annual Review of Materials Research, 2007. **37**(1): p. 93-126.

72. Gianola, D.S. and C. Eberl, *Micro- and nanoscale tensile testing of materials.* JOM, 2009. **61**(3): p. 24-35.

73. Haque, M.A. and M.T.A. Saif, *A review of MEMS-based microscale and nanoscale tensile and bending testing.* Experimental Mechanics, 2003. **43**(3): p. 248-255.

74. Mine, Y., H. Takashima, M. Matsuda, and K. Takashima, *Microtension behaviour of lenticular martensite structure of Fe–30 mass% Ni alloy.* Materials Science and Engineering: A, 2014. **618**: p. 359-367.

75. Slaby, S.A., O. Kraft, and C. Eberl, *Fatigue properties of conventionally manufactured and micro-powder-injection-moulded 17-4PH micro-components.* Fatigue & Fracture of Engineering Materials & Structures, 2016. **39**(6): p. 780-789.

76. Sato, K., T. Yoshioka, T. Ando, M. Shikida, and T. Kawabata, *Tensile testing of silicon film having different crystallographic orientations carried out on a silicon chip.* Sensors and Actuators A: Physical, 1998. **70**(1): p. 148-152.

77. Ensslen, C., O. Kraft, R. Mönig, J. Xu, G.-P. Zhang, and R. Schneider, *Mechanical annealing of Cu–Si nanowires during high-cycle fatigue.* MRS Communications, 2014. **4**(3): p. 83-87.

78. Ensslen, C., C. Brandl, G. Richter, R. Schwaiger, and O. Kraft, *Notch insensitive strength and ductility in gold nanowires.* Acta Materialia, 2016. **108**: p. 317-324.

79. *In situ nanomechanical testing in focused ion beam and scanning electron microscopes.* Review of Scientific Instruments, 2011. **82**(6): p. 063901.

80. Mine, Y., K. Hirashita, M. Matsuda, M. Otsu, and K. Takashima, *Effect of hydrogen on tensile behaviour of micrometre-sized specimen fabricated from a metastable austenitic stainless steel.* Corrosion Science, 2011. **53**(2): p. 529-533.

81. Mine, Y., K. Hirashita, H. Takashima, M. Matsuda, and K. Takashima, *Micro-tension behaviour of lath martensite structures of carbon steel.* Materials Science and Engineering: A, 2013. **560**: p. 535-544.

82. Mine, Y., H. Fujisaki, M. Matsuda, M. Takeyama, and K. Takashima, *Microtension behaviour of TiAl polysynthetically twinned crystals with 0°- and 90°-oriented lamellae.* Scripta Materialia, 2011. **65**(8): p. 707-710.

83. Higo, Y., K. Takashima, M. Shimojo, S. Sugiura, B. Pfister, and M.V. Swain, *Fatigue Testing Machine of Micro-Sized Specimens for MEMS Applications.* MRS Proceedings, 1999. **605**.

84. Orso, S., U.G.K. Wegst, C. Eberl, and E. Arzt, *Micrometer-Scale Tensile Testing of Biological Attachment Devices.* Advanced Materials, 2006. **18**(7): p. 874-877.

85. Kiener, D., W. Grosinger, G. Dehm, and R. Pippan, *A further step towards an understanding of size-dependent crystal plasticity: In situ tension experiments of miniaturized single-crystal copper samples.* Acta Materialia, 2008. **56**(3): p. 580-592.

86. Irwin, G.R., *Analysis of Stresses and Strains Near the End of a Crack Traversing a Plate.* J. Appl. Mech., 1957.

87. Rice, J.R., *A Path Independent Integral and the Approximate Analysis of Strain Concentration by Notches and Cracks.* Journal of Applied Mechanics, 1968. **35**(2): p. 379-386.

88. *Standard Test Methods for Determination of Fracture Toughness of Advanced Ceramics at Ambient Temperature*, 2015, ASTM International.

89. e.V., D.D.I.f.N., *Hochleistungskeramik - Prüfverfahren zur Bestimmung der Bruchzähigkeit von monolithischer Keramik - Teil 5: Verfahren für Biegeproben mit V-Kerb (SEVNB-Verfahren)*, 2004, Beuth Verlag GmbH. p. 20.

90. Shih, T.T. and J. Opoku, *Application of fracture mechanics to ceramic materials—A state-of-the-art review.* Engineering Fracture Mechanics, 1979. **12**(4): p. 479-498.

91. Zhu, X.-K. and J.A. Joyce, *Review of fracture toughness (G, K, J, CTOD, CTOA) testing and standardization.* Engineering Fracture Mechanics, 2012. **85**: p. 1-46.

92. Liu, S., J.M. Wheeler, P.R. Howie, X.T. Zeng, J. Michler, and W.J. Clegg, *Measuring the fracture resistance of hard coatings.* Applied Physics Letters, 2013. **102**(17): p. 171907.

93. Sebastiani, M., K.E. Johanns, E.G. Herbert, F. Carassiti, and G.M. Pharr, *A novel pillar indentation splitting test for measuring fracture toughness of thin ceramic coatings.* Philosophical Magazine, 2015. **95**(16-18): p. 1928-1944.

94. Sokalski, K., Jankowski, B. and Ślusarek, B. , *Binary Relations between Magnitudes of Different Dimensions Used in Material Science Optimization Problems Pseudo-State Equation of Soft Magnetic Composites.* Materials Sciences and Applications, 2014. **5**: p. 1040-1047.

95. Höganäs. *Somaloy Material Data.* [PDF] 2016 [cited 2017 22.05.2017]; Available from: https://www.hoganas.com/en/business-areas/soft-magnetic-composites/downloads/.

96. Kim, K., M. Watanabe, S. Kuroda, and N. Kawano, *Observation of High Resolution Microstructures in Thermal Sprayed Coatings and Single Deposited Splats Using Ion Beam Milling.* Materials Transactions, JIM, 2011. **52**(3): p. 439-446.

97. N. Erdman, R.C., and S. Asahina, *Precise SEM Cross Section Polishing via Argon Beam Milling.* Microscopy Today, 2006.

98. Shunsuke Asahina, F.C., Sam M. Stevens and Osamu Terasaki, *An application to Ultra High Resolution SEM of Broad Argon Ion Beam Cross-sectional technique.* MC 2009, Instrumentation and Methodology, 2009. **1**.

99. Radi, Z., K. Havancs, x00E, S. Kal, cska, and A. Baris. *Surface polishing and slope cut by parallel ar ion beams for high resolution electron backscatter diffraction measurements*. in *2014 International Conference on Nanoscience and Nanotechnology*. 2014.

100. Carter, G., M.J. Nobes, and I.V. Katardjiev, *The theory of ion beam polishing and machining.* Vacuum, 1993. **44**(3): p. 303-309.

101. Oettel, H. and H. Schumann, *Metallografie: mit einer Einführung in die Keramografie*2011: Wiley-VCH-Verlag.

102. *Standard Test Method for Determining Volume Fraction by Systematic Manual Point Count*, 2011, ASTM International.

103. Lekstrom, M., M.A. McLachlan, S. Husain, D.W. McComb, and B.A. Shollock, *Using the in situ lift-out technique to prepare TEM specimens on a single-beam FIB instrument.* Journal of Physics: Conference Series, 2008. **126**(1): p. 012028.

104. Zhang, H., B.E. Schuster, Q. Wei, and K.T. Ramesh, *The design of accurate micro-compression experiments.* Scripta Materialia, 2006. **54**(2): p. 181-186.

105. Schwaiger, R., M. Weber, B. Moser, P. Gumbsch, and O. Kraft, *Mechanical assessment of ultrafine-grained nickel by microcompression experiment and finite element simulation.* Journal of Materials Research, 2012. **27**(01): p. 266-277.

106. Schwark, T., O. Kraft, and R. Schwaiger, *The boundaries of soft magnetic composites reveal their complexity in compression and bending tests at the micro-scale.* Materials Science and Engineering: A, 2017. **684**: p. 270-274.

107. Singh, D.R.P., N. Chawla, G. Tang, and Y.L. Shen, *Micropillar compression of Al/SiC nanolaminates.* Acta Materialia, 2010. **58**(20): p. 6628-6636.

108. In-suk Choi, Y.G., Daniel Kaufmann, Oliver Kraft, Ruth Schwaiger, *Measurement of Young's modulus of anisotropic materials using microcompression testing.* Journal of Materials Research 2012. **27**(21): p. 2752 – 2759.

109. Gong, J. and A.J. Wilkinson, *Anisotropy in the plastic flow properties of single-crystal α titanium determined from micro-cantilever beams.* Acta Materialia, 2009. **57**(19): p. 5693-5705.

110. Weihs, T.P., S. Hong, J.C. Bravman, and W.D. Nix, *Mechanical deflection of cantilever microbeams: A new technique for testing the mechanical properties of thin films.* Journal of Materials Research, 1988. **3**(05): p. 931-942.

111. DIN50115:1991-04, *Notched bar impact testing of metallic materials using test pieces other than ISO test pieces*, 1991, Beuth Verlag GmbH.

112. Nishida, T., Y. Hanaki, and G. Pezzotti, *Effect of Notch-Root Radius on the Fracture Toughness of a Fine-Grained Alumina.* Journal of the American Ceramic Society, 1994. **77**(2): p. 606-608.

113. Degnan, C.C., A.R. Kennedy, and P.H. Shipway, *Fracture toughness measurements of powder metallurgical (P/M) green compacts: A novel method of sample preparation.* Journal of Materials Science, 2004. **39**(7): p. 2605-2607.

114. Roesler, J., H. Harders, and M. Baeker, *Mechanical Behaviour of Engineering Materials: Metals, Ceramics, Polymers, and Composites* 2007: Springer Berlin Heidelberg.

115. Greer, J.R., W.C. Oliver, and W.D. Nix, *Size dependence of mechanical properties of gold at the micron scale in the absence of strain gradients.* Acta Materialia, 2005. **53**(6): p. 1821-1830.

116. Ledbetter, H.M. and R.P. Reed, *Elastic Properties of Metals and Alloys, I. Iron, Nickel, and Iron - Nickel Alloys.* Journal of Physical and Chemical Reference Data, 1973. **2**(3): p. 531-618.

117. R. Bures, e.a., *Structure and Properties of Composites Based on Mixed Morphology of Ferromagnetic Particles.* Acta Phys. Pol. A, 2014. **126**: p. 140-141.

118. Schreijäg, S., D. Kaufmann, M. Wenk, O. Kraft, and R. Mönig, *Size and microstructural effects in the mechanical response of α-Fe and low alloyed steel.* Acta Materialia, 2015. **97**: p. 94-104.

119. Motz, C., D. Weygand, J. Senger, and P. Gumbsch, *Micro-bending tests: A comparison between three-dimensional discrete dislocation dynamics simulations and experiments.* Acta Materialia, 2008. **56**(9): p. 1942-1955.

120. Jin, Z.H. and R.C. Batra, *Some basic fracture mechanics concepts in functionally graded materials.* Journal of the Mechanics and Physics of Solids, 1996. **44**(8): p. 1221-1235.

121. Syurik, J., R. Schwaiger, P. Sudera, S. Weyand, S. Johnsen, G. Wiegand, and H. Hölscher, *Bio-inspired micro-to-nanoporous polymers with tunable stiffness.* Beilstein Journal of Nanotechnology, 2017. **8**: p. 906-914.

122. Gao, H., B. Ji, I.L. Jäger, E. Arzt, and P. Fratzl, *Materials become insensitive to flaws at nanoscale: Lessons from nature.* Proceedings of the National Academy of Sciences, 2003. **100**(10): p. 5597-5600.

123. Hodge, A.M., J. Biener, J.R. Hayes, P.M. Bythrow, C.A. Volkert, and A.V. Hamza, *Scaling equation for yield strength of nanoporous open-cell foams.* Acta Materialia, 2007. **55**(4): p. 1343-1349.

124. Shih, C., Y. Katoh, K.J. Leonard, H. Bei, and E. Lara-Curzio, *Determination of interfacial mechanical properties of ceramic composites by the compression of micro-pillar test specimens.* Journal of Materials Science, 2013. **48**(15): p. 5219-5224.

125. Meza, J.M., M.C.M. Farias, R.M.d. Souza, and L.J.C. Riaño, *Using the ratio: maximum load over unload stiffness squared, Pm/Su2, on the evaluation of machine stiffness and area function of blunt indenters on depth-sensing indentation equipment.* Materials Research, 2007. **10**: p. 437-447.

126. Pharr, G.M., J.H. Strader, and W.C. Oliver, *Critical issues in making small-depth mechanical property measurements by nanoindentation with continuous stiffness measurement.* Journal of Materials Research, 2009. **24**(3): p. 653-666.

127. Cordill, M.J., N.R. Moody, and W.W. Gerberich, *Effects of dynamic indentation on the mechanical response of materials.* Journal of Materials Research, 2008. **23**(6): p. 1604-1613.

Schriftenreihe des Instituts für Angewandte Materialien

ISSN 2192-9963

Band 1 Prachai Norajitra
Divertor Development for a Future Fusion Power Plant. 2011
ISBN 978-3-86644-738-7

Band 2 Jürgen Prokop
Entwicklung von Spritzgießsonderverfahren zur Herstellung von Mikrobauteilen durch galvanische Replikation. 2011
ISBN 978-3-86644-755-4

Band 3 Theo Fett
New contributions to R-curves and bridging stresses – Applications of weight functions. 2012
ISBN 978-3-86644-836-0

Band 4 Jérôme Acker
Einfluss des Alkali/Niob-Verhältnisses und der Kupferdotierung auf das Sinterverhalten, die Strukturbildung und die Mikrostruktur von bleifreier Piezokeramik $(K_{0,5}Na_{0,5})NbO_3$. 2012
ISBN 978-3-86644-867-4

Band 5 Holger Schwaab
Nichtlineare Modellierung von Ferroelektrika unter Berücksichtigung der elektrischen Leitfähigkeit. 2012
ISBN 978-3-86644-869-8

Band 6 Christian Dethloff
Modeling of Helium Bubble Nucleation and Growth in Neutron Irradiated RAFM Steels. 2012
ISBN 978-3-86644-901-5

Band 7 Jens Reiser
Duktilisierung von Wolfram. Synthese, Analyse und Charakterisierung von Wolframlaminaten aus Wolframfolie. 2012
ISBN 978-3-86644-902-2

Band 8 Andreas Sedlmayr
Experimental Investigations of Deformation Pathways in Nanowires. 2012
ISBN 978-3-86644-905-3

Band 9 Matthias Friedrich Funk
Microstructural stability of nanostructured fcc metals during cyclic deformation and fatigue. 2012
ISBN 978-3-86644-918-3

Band 10 Maximilian Schwenk
Entwicklung und Validierung eines numerischen Simulationsmodells zur Beschreibung der induktiven Ein- und Zweifrequenzrandschichthärtung am Beispiel von vergütetem 42CrMo4. 2012
ISBN 978-3-86644-929-9

Band 11 Matthias Merzkirch
Verformungs- und Schädigungsverhalten der verbundstranggepressten, federstahldrahtverstärkten Aluminiumlegierung EN AW-6082. 2012
ISBN 978-3-86644-933-6

Band 12 Thilo Hammers
Wärmebehandlung und Recken von verbundstranggepressten Luftfahrtprofilen. 2013
ISBN 978-3-86644-947-3

Band 13 Jochen Lohmiller
Investigation of deformation mechanisms in nanocrystalline metals and alloys by in situ synchrotron X-ray diffraction. 2013
ISBN 978-3-86644-962-6

Band 14 Simone Schreijäg
Microstructure and Mechanical Behavior of Deep Drawing DC04 Steel at Different Length Scales. 2013
ISBN 978-3-86644-967-1

Band 15 Zhiming Chen
Modelling the plastic deformation of iron. 2013
ISBN 978-3-86644-968-8

Band 16 Abdullah Fatih Çetinel
Oberflächendefektausheilung und Festigkeitssteigerung von niederdruckspritzgegossenen Mikrobiegebalken aus Zirkoniumdioxid. 2013
ISBN 978-3-86644-976-3

Band 17 Thomas Weber
Entwicklung und Optimierung von gradierten Wolfram/EUROFER97-Verbindungen für Divertorkomponenten. 2013
ISBN 978-3-86644-993-0

Die Bände sind unter www.ksp.kit.edu als PDF frei verfügbar oder als Druckausgabe bestellbar.

Band 18 Melanie Senn
Optimale Prozessführung mit merkmalsbasierter Zustandsverfolgung. 2013
ISBN 978-3-7315-0004-9

Band 19 Christian Mennerich
Phase-field modeling of multi-domain evolution in ferromagnetic shape memory alloys and of polycrystalline thin film growth. 2013
ISBN 978-3-7315-0009-4

Band 20 Spyridon Korres
On-Line Topographic Measurements of Lubricated Metallic Sliding Surfaces. 2013
ISBN 978-3-7315-0017-9

Band 21 Abhik Narayan Choudhury
Quantitative phase-field model for phase transformations in multi-component alloys. 2013
ISBN 978-3-7315-0020-9

Band 22 Oliver Ulrich
Isothermes und thermisch-mechanisches Ermüdungsverhalten von Verbundwerkstoffen mit Durchdringungsgefüge (Preform-MMCs). 2013
ISBN 978-3-7315-0024-7

Band 23 Sofie Burger
High Cycle Fatigue of Al and Cu Thin Films by a Novel High-Throughput Method. 2013
ISBN 978-3-7315-0025-4

Band 24 Michael Teutsch
Entwicklung von elektrochemisch abgeschiedenem LIGA-Ni-Al für Hochtemperatur-MEMS-Anwendungen. 2013
ISBN 978-3-7315-0026-1

Band 25 Wolfgang Rheinheimer
Zur Grenzflächenanisotropie von $SrTiO_3$. 2013
ISBN 978-3-7315-0027-8

Band 26 Ying Chen
Deformation Behavior of Thin Metallic Wires under Tensile and Torsional Loadings. 2013
ISBN 978-3-7315-0049-0

Band 27 Sascha Haller
Gestaltfindung: Untersuchungen zur Kraftkegelmethode. 2013
ISBN 978-3-7315-0050-6

Band 28 Nicht erschienen

Band 29 Gunnar Picht
Einfluss der Korngröße auf ferroelektrische Eigenschaften dotierter $Pb(Zr_{1-x}Ti_x)O_3$ Materialien. 2013
ISBN 978-3-7315-0106-0

Band 30 Esther Held
Eigenspannungsanalyse an Schichtverbunden mittels inkrementeller Bohrlochmethode. 2013
ISBN 978-3-7315-0127-5

Band 31 Pei He
On the structure-property correlation and the evolution of Nanofeatures in 12-13.5% Cr oxide dispersion strengthened ferritic steels. 2014
ISBN 978-3-7315-0141-1

Band 32 Jan Hoffmann
Ferritische ODS-Stähle – Herstellung, Umformung und Strukturanalyse. 2014
ISBN 978-3-7315-0157-2

Band 33 Wiebke Sittel
Entwicklung und Optimierung des Diffusionsschweißens von ODS Legierungen. 2014
ISBN 978-3-7315-0182-4

Band 34 Osama Khalil
Isothermes Kurzzeitermüdungsverhalten der hoch-warmfesten Aluminium-Knetlegierung 2618A (AlCu2Mg1,5Ni). 2014
ISBN 978-3-7315-0208-1

Band 35 Nicht erschienen

Band 36 Christoph Hage
Grundlegende Aspekte des 2K-Metallpulverspritzgießens. 2014
ISBN 978-3-7315-0217-3

Band 37 Bartłomiej Albiński
Instrumentierte Eindringprüfung bei Hochtemperatur für die Charakterisierung bestrahlter Materialien. 2014
ISBN 978-3-7315-0221-0

Band 38 Tim Feser
Untersuchungen zum Einlaufverhalten binärer alpha-Messinglegierungen unter Ölschmierung in Abhängigkeit des Zinkgehaltes. 2014
ISBN 978-3-7315-0224-1

Band 39 Jörg Ettrich
Fluid Flow and Heat Transfer in Cellular Solids. 2014
ISBN 978-3-7315-0241-8

Band 40 Melanie Syha
Microstructure evolution in strontium titanate Investigated by means of grain growth simulations and x-ray diffraction contrast tomography experiments. 2014
ISBN 978-3-7315-0242-5

Band 41 Thomas Haas
Mechanische Zuverlässigkeit von gedruckten und gasförmig abgeschiedenen Schichten auf flexiblem Substrat. 2014
ISBN 978-3-7315-0250-0

Band 42 Aron Kneer
Numerische Untersuchung des Wärmeübertragungsverhaltens in unterschiedlichen porösen Medien. 2014
ISBN 978-3-7315-0252-4

Band 43 Manuel Feuchter
Investigations on Joule heating applications by multiphysical continuum simulations in nanoscale systems. 2014
ISBN 978-3-7315-0261-6

Band 44 Alexander Vondrous
Grain growth behavior and efficient large scale simulations of recrystallization with the phase-field method. 2014
ISBN 978-3-7315-0280-7

Band 45 Tobias Kennerknecht
Fatigue of Micro Molded Materials – Aluminum Bronze and Yttria Stabilized Zirconia. 2014
ISBN 978-3-7315-0293-7

Band 46 Christopher Scherr
Elektrochemisches Verhalten von Lithium-Schwefel-Zellen mit unterschiedlicher Kathodenstruktur. 2015
ISBN 978-3-7315-0296-8

Die Bände sind unter www.ksp.kit.edu als PDF frei verfügbar oder als Druckausgabe bestellbar.

Band 47 Konstantin Frölich
Der Decal-Prozess zur Herstellung katalysatorbeschichteter Membranen für PEM-Brennstoffzellen. 2015
ISBN 978-3-7315-0334-7

Band 48 Benedikt Haspel
Werkstoffanalytische Betrachtung der Eigenschaften von mittels neuartiger RTM-Fertigungsprozesse hergestellten glasfaserverstärkten Polymerverbunden. 2015
ISBN 978-3-7315-0337-8

Band 49 Marco Berghoff
Skalenübergreifende Modellierung und Optimierung vom atomistischen kristallinen Phasenfeldmodell bis zur mesoskopischen Phasenfeldmethode. 2015
ISBN 978-3-7315-0416-0

Band 50 Michael Selzer
Mechanische und Strömungsmechanische Topologieoptimierung mit der Phasenfeldmethode. 2016
ISBN 978-3-7315-0431-3

Band 51 Michael Mahler
Entwicklung einer Auswertemethode für bruchmechanische Versuche an kleinen Proben auf der Basis eines Kohäsivzonenmodells. 2016
ISBN 978-3-7315-0441-2

Band 52 Christoph Bohnert
Numerische Untersuchung des Verformungs- und Bruchverhaltens von einkristallinem Wolfram auf mikroskopischer Ebene. 2016
ISBN 978-3-7315-0444-3

Band 53 Stefan Guth
Schädigung und Lebensdauer von Nickelbasislegierungen unter thermisch-mechanischer Ermüdungsbeanspruchung bei verschiedenen Phasenlagen. 2016
ISBN 978-3-7315-0445-0

Band 54 Markus Klinsmann
The Effects of Internal Stress and Lithium Transport on Fracture in Storage Materials in Lithium-Ion Batteries. 2016
ISBN 978-3-7315-0455-9

Band 55 Thomas Straub
Experimental Investigation of Crack Initiation in Face-Centered Cubic Materials in the High and Very High Cycle Fatigue Regime. 2016
ISBN 978-3-7315-0471-9

Band 56 Maren Lepple
Kupfer- und Eisenoxide als Konversions-Elektrodenmaterialien für Lithium-Ionen-Batterien: Thermodynamische und Elektrochemische Untersuchungen. 2016
ISBN 978-3-7315-0482-5

Band 57 Stefan Andreas Slaby
Charakterisierung und Bewertung der Zug- und Ermüdungseigenschaften von Mikrobauteilen aus 17-4PH Edelstahl. Ein Vergleich von mikropulverspritzgegossenem und konventionell hergestelltem Material. 2017
ISBN 978-3-7315-0484-9

Band 58 Kumar Ankit
Phase-field modeling of microstructural pattern formation in alloys and geological veins. 2016
ISBN 978-3-7315-0491-7

Band 59 Kuo Zhang
Characterization and Modeling of the Ratcheting Behavior of the Ferritic-Martensitic Steel P91. 2017
ISBN 978-3-7315-0503-7

Band 60 Nicht erschienen

Band 61 Fabian Lemke
Untersuchung des Sinterverhaltens von $SrTiO_3$ unter Berücksichtigung der Defektchemie. 2016
ISBN 978-3-7315-0510-5

Band 62 Johannes Kümmel
Detaillierte Analyse der Aufbauschneidenbildung bei der Trockenzerspanung von Stahl C45E mit Berücksichtigung des Werkzeugverschleißes. 2016
ISBN 978-3-7315-0518-1

Band 63 László Hagymási
Modellierung der Stoffübertragung beim Niederdruckcarbonitrieren mit Ammoniak und Acetylen. 2016
ISBN 978-3-7315-0568-6

Band 64 Reza Eslami
A novel micro-mechanical model for prediction of multiaxial high cycle fatigue at small scales. 2017
ISBN 978-3-7315-0583-9

Band 65 Sebastian Schulz
Phase-field simulations of multi-component solidification and coarsening based on thermodynamic datasets. 2017
ISBN 978-3-7315-0618-8

Band 66 Markus Stricker
Die Übertragung von mikrostrukturellen Eigenschaften aus der diskreten Versetzungsdynamik in Kontinuumsbeschreibungen. 2017
ISBN 978-3-7315-0658-4

Band 67 Luis Straßberger
Untersuchung und Modellierung des viskoplastischen Verformungsverhaltens oxidpartikelverstärkter Stähle. 2018
ISBN 978-3-7315-0674-4

Band 68 Mark Wobrock
Microplasticity of idealized single crystalline Ag cantilevers characterized with methods of high resolution. 2017
ISBN 978-3-7315-0682-9

Band 69 Amritesh Kumar
Micromechanical study on the deformation behaviour of directionally solidified NiAl-Cr eutectic composites. 2017
ISBN 978-3-7315-0694-2

Band 70 Johannes Hötzer
Massiv-parallele und großskalige Phasenfeldsimulationen zur Untersuchung der Mikrostrukturentwicklung. 2017
ISBN 978-3-7315-0693-5

Band 71 Thomas Hupfer
Herstellung von LATP für den Einsatz als Festkörperelektrolyt und dessen Eigenschaften. 2017
ISBN 978-3-7315-0702-4

Band 72 Florentin Pottmeyer
Schädigungsverhalten von in CFK-Laminaten eingebetteten Inserts unter bauteilnahen Beanspruchungen. 2017
ISBN 978-3-7315-0719-2

Band 73 Andres Höweling
Untersuchung der Hochvoltstabilität und Tiefentladung von dotierten $LiNi_{0,5}Mn_{1,5}O_4$-Hochvoltspinellen. 2018
ISBN 978-3-7315-0728-4

Band 74 Tabea Gisela Schwark
Deformation and Fracture Properties of the Soft Magnetic Composite Somaloy 700 3P on Different Length Scales. 2018
ISBN 978-3-7315-0759-8